Studies in Infrastructure and Control

The book series aims to publish top-quality state-of-the-art textbooks, research monographs, edited volumes and selected conference proceedings related to infrastructure, innovation, control, and related fields. Additionally, established and emerging applications related to applied areas like smart cities, internet of things, machine learning, artificial intelligence, etc., are developed and utilized in an effort to demonstrate recent innovations in infrastructure and the possible implications of control theory therein. The study also includes areas like transportation infrastructure, building infrastructure management and seismic vibration control, and also spans a gamut of areas from renewable energy infrastructure like solar parks, wind farms, biomass power plants and related technologies, to the associated policies and related innovations and control methodologies involved.

More information about this series at https://link.springer.com/bookseries/16625

Anita Khosla · Monika Aggarwal
Editors

Smart Structures in Energy Infrastructure

Proceedings of ICRTE 2021, Volume 2

Springer

Editors
Anita Khosla
Department of Electrical and Electronics Engineering
Faculty of Engineering and Technology
Manav Rachna International Institute of Research and Studies
Faridabad, Haryana, India

Monika Aggarwal
Centre for Applied Research in Electronics
Indian Institute of Technology
New Delhi, Delhi, India

ISSN 2730-6453 ISSN 2730-6461 (electronic)
Studies in Infrastructure and Control
ISBN 978-981-16-4743-7 ISBN 978-981-16-4744-4 (eBook)
https://doi.org/10.1007/978-981-16-4744-4

This Springer imprint is published by the registered company Springer Nature Singapore Pte Ltd.
The registered company address is: 152 Beach Road, #21-01/04 Gateway East, Singapore 189721, Singapore

Dedicated

to

Almighty

Preface

It is a matter of great pleasure and honour that we had organized the prestigious International Conference on Renewable Technologies in Engineering. (ICRTE 2021), in association with Springer in Natureon 15–16 April 2021 at Manav Rachna International Institute of Research and Studies, Faridabad, India, in virtual mode due to the ongoing pandemic situation. The aim of this conference is to bring together academicians, researchers, scientists, engineers, and practitioners to exchange and share their experiences, ideas, and the results in the area of renewable sources integration planning and control and their optimization solutions, smart structures for intelligent power with use of energy storage and transportation systems, and industry innovation and automation in smart structures

Renewable energy is useful energy that is collected from renewable resources, which are naturally replenished on a human timescale, including carbon neutral sources like sunlight, wind, rain, tides, waves, and geothermal heat. It is often referred to as clean energy. It is with very idea in mind, the Department of Electrical, Electronics and Communication Engineering, Manav Rachna International Institute of Research and Studies, Faridabad, in association with Springer had taken this initiative in organizing the International Conference on Renewable Technologies in Engineering. (ICRTE 2021), on 15–16 April 2021.

During the conference, research scholars, academicians, and industry experts, from different part of globe, deliberated for two days on the interdisciplinary areas like renewable sources integration planning and control and their optimization solutions, smart structures for intelligent power with use of energy storage and transportation systems and industry innovation and automation in smart structures.

It is great pleasure and honour for us to bring the proceedings of the International Conference on Renewable Technologies in Engineering (ICRTE 2021). All presented papers are arranged in two books.

Renewable Energy Optimization, Planning and Control & Smart Structures in Energy Infrastructure.

ICRTE 2021 is our humble effort at establishing a base, platform, against which we can build a brighter tomorrow.

Faridabad, India Dr. Anita Khosla
New Delhi, India Dr. Monika Aggarwal

Acknowledgements

We would like to express our gratitude to people without whom support we would have not been able to conduct this conference and compile all the papers.

We are pleased to note the overwhelming response to our call of papers from the authors for the conference. Our sincere thanks to all the authors for their contributions. We are also thankful to International Advisory Committee, National Advisory Committee, Local Advisory Committee, and Springer for their motivation and support. We would like to extend our gratitude to reviewers who have worked hard in selecting high-quality papers and writing review reports.

Our special thanks to dignitaries gracing the inaugural function and the keynote speakers. We would also like to thank the session chairs for their valuable support in the smooth conduct of technical sessions. We are also thankful to our management for guidance and liberal help for smooth conduct and success of conference.

We are happy to mention that organizing team of ICRTE 2021 has done a professional exercise in selecting the research papers and arranging them in technical session.

Dr. Anita Khosla
Dr. Monika Aggarwal

Contents

1 **ANN Model-Based Performance Simulation of a Solar PV Operated Helical Rotor Water Pump** 1
Yuvraj Praveen Soni and E. Fernandez

2 **A Comparative Analysis of Neural Network-Based Models for Forecasting of Solar Irradiation with Different Learning Algorithms** 9
Anuj Gupta, Kapil Gupta, and Sumit Saroha

3 **A Review of Load Frequency Control of Hybrid Power System** 19
Amit Atri and Anita Khosla

4 **Time Duration Prediction of Electrical Power Outages** 31
Rishabh Doshi, Rishabh Dev Saini, and Shivam Kansal

5 **Comparative THD Analysis of Multi-level Inverter Using SPWM Scheme** 41
Shashi Shekhar Tripathi, Manoj Kumar Kar, and A. K. Singh

6 **Power Quality Improvement of Railway Traction System Using D-STATCOM** 49
Ruma Sinha and H. A. Vidya

7 **Mitigation of Harmonics Using Passive-Series Active-Hybrid Filter in 1-Φ and 3-Φ System Feeding Nonlinear Load** 59
G. Jayachitra, H. A. Vidya, and Ruma Sinha

8 **Thermal Modelling of Solar Photovoltaic Panel Using FEM** 69
Abhilash Narasimhan

9 **Selective Harmonic Elimination for Cascade Multilevel Inverter Using Genetic Algorithm** 77
Hemant Gupta, Arvind Yadav, and Sanjay Maurya

10 A Concept Design of a Futuristic Battery Management System for Submarines Using IEEE802.3bt Network 87
Arun Singh and Anita Khosla

11 Technological Advancements for Reduced Charging Time of Electric Vehicle Batteries: A Review 99
Abdullah Naim and Devendra Vashist

12 Compact and Efficient Way of LSEV Charging 113
Rakesh Sharma and Anita Khosla

13 Optimization of Closed Loop Controlled Charging Time of Li-Ion Battery Using ANFIS 123
Prakash Bahrani and Naveen Jain

14 Electric Vehicle Reliability Assessment Based on Fault Tree Analysis 135
Kailash Rana and Dheeraj Joshi

15 Design Analysis of Dimmer Light for Autonomous Vehicles 145
Abhisheak Mangla, Dhruv Gulati, Nanak Jhamb, and Devendra Vashist

16 Comparative Analysis of High-Gain Transformerless DC–DC Converter for DC Mircogrids 153
K. V. Suresh, K. U. Vinayaka, and N. V. Jyothi

17 Energy Harvesting with Photovoltaic Arrays: Assessment of Reliability with Alternative Configurations for Power Delivery 167
Sandhya Prajapati, Yuvraj Praveen Soni, and E. Fernandez

18 A Sensors-Based Solar-Powered Smart Irrigation System Using IoT 175
Sandeep Chopade, Swati Chopade, and Sushopti Gawade

19 Drone Development and Embellishing It into Crop Monitoring and Protection Along with Pesticide Spraying Mechanism 199
Smita Agrawal, Preeti Kathiria, Vishwam Rawal, and Trushit Vyas

Author Index 211

About the Editors

Dr. Anita Khosla is working as Professor in the Department of Electrical and Electronics Engineering at Faculty of Engineering and Technology, Manav Rachna International Institute of Research and Studies, University Faridabad (Haryana). She completed B.Tech. in Electrical Engineering from National Institute of Technology, Kurukshetra. She did her M.Tech. in ECE and Ph.D. in Control Systems (Electrical Engineering) from Manav Rachna International Institute of Research and Studies, University Faridabad (Haryana). She has a rich experience of 26 years in teaching. She has published 40 research papers in National, International Journals of high repute and conferences. She is author of three books, and one more is under progress. She has attended and organized many workshops, FDPs and seminars. She has been a session speaker also. She is member of technical societies—Institution of Engineers and ISTE. She was Reviewer of book chapters for Pearson Publishers.

Dr. Monika Aggarwal is working as Professor in Centre for Applied Research in Electronics—Indian Institute of Technology, Delhi. She received the B.Tech. degree in Electrical Engineering and the M.Tech. degree in Electronics and Communication Engineering from the Regional Engineering College, Kurukshetra, India, in 1993 and 1995. She has done Ph.D. degree with Specialization in Signal Processing and Communications from the Indian Institute of Technology, New Delhi, India, in 2000, respectively. She was employed with Hughes Software Systems (HSS), Gurgaon, India, from 1999 to 2002. During 2001, she was a Visiting Researcher in the Department of Systems and Control, Uppsala University, Uppsala, Sweden. She has published 62 research papers in National, International Journals of high repute and conferences and was also involved in funded research projects. She has been delivered many expert lectures on signal processing.

Chapter 1
ANN Model-Based Performance Simulation of a Solar PV Operated Helical Rotor Water Pump

Yuvraj Praveen Soni and E. Fernandez

Abstract PV power can be best utilized for water pumping systems (WPS) for irrigation and drinking water applications. For the small DC application, generally permanent magnet DC motor (PMDCM) is preferred for direct coupling with the PV system. This is because of its higher efficiency, moderate cost, high reliability, good starting torque, and operation over a wide range of input voltages. This work focuses on the development of efficient tool which can be utilized to analyze the system performance in terms of efficiency, average flow, motor speed, and power output for which feedforward backpropagation artificial neural network (ANN) is developed. Impact on WPS on varying pump head is studied and observed that head can be chosen ranging from 30 to 100% of its base value for good performance.

Keywords Solar system · Water pumping system · ANN · Helical rotor · DC motor

1.1 Introduction

PV is seen as a valuable source for water pumping system (WPS), which has gained enormous popularity in supplying water for domestic and irrigation purposes with great acceptance and decent reliability. PV generates DC power, which makes it simple to get easily connected with the DC motor. Brushless DC motor is the most common choice for WPS. It incorporates numerous advantages such as salient operation with low maintenance, high reliability, compact size, and high efficiency. A DC motor is preferred because it can be directly connected without any complex intermediate system, thus reducing the overall implementation cost.

Posorki [1] demonstrated the PV based WPS in seven different countries. They deployed 90 systems with a rated capacity of 180 kWp. The conclusion is drawn that the PV based WPS exhibits a more cost-efficient system than diesel-based WPS. Further, the comparison between induction motor (IM) and DC motor is shown in the

Y. P. Soni (✉) · E. Fernandez
Indian Institute of Technology Roorkee, Roorkee, India

A. Khosla and M. Aggarwal (eds.), *Smart Structures in Energy Infrastructure*,
Studies in Infrastructure and Control,
https://doi.org/10.1007/978-981-16-4744-4_1

literature [2–4] to investigate the performance in the presence of maximum power point tracking (MPPT) system. It was observed that for all values of irradiance, the DC motor performed much better than IM. Chandrasekaran et al. [5] compared conventional motor and PMDCM. The analysis showed that, comparatively, PMDCM is more efficient.

Vick and Clark [6] examine WPS with helical rotor at three different head of 50, 75, and 100 m. Two systems rated 480 and 640 Wp are analyzed and conclude that satisfactory performance is observed in the system. Further, the proposed method was economical, which can be brought in utilization for farmers and ranchers in developing countries. Literature [7] investigated the WPS deployed for livestock watering and agricultural purposes. The system contains a helical rotor pump connected with brushless DC motor.

Biji [8] implemented the ANN technique to operate the PV system at maximum power operating point. Also, ANN has been explored in numerous dimensions of the electrical sector such as fault detection [9], load flow [10], controlling [11], and energy forecasting [12].

Table 1.1 shows the comparison table of literature's investigated in analyzing the performance of the WPS system.

Numerous research has been done on the performance analysis of WPS to examine the volumetric water flow. Most of the studies are confined to an analysis of the system with changes in PV parameters and average flow. In this paper, study has been done to analyze the performance of WPS, evaluating efficiency, average water flow, motor speed, and power output with the change in head of the water pump through developed feedforward ANN tool. The paper is organized as follows—Sect. 1.2 discusses the WPS considered for study. Section 1.3 discusses the developed ANN model to examine the system's performance, followed by results and discussion in Sect. 1.4. The conclusion is deliberated in Sect. 1.5.

Table 1.1 Literature's comparison table

References	System description	Input parameter	Performance parameter
[13]	PV-WPS with low head	Irradiance, temperature, tension, and current	Flow rate
[14]	Directly PV powered DC PM motor—propeller system	PV array size	Thrust, power output, and efficiency
[15]	PV-WPS	Irradiance	Power and volume of water flow
[16]	Solar photovoltaic irrigation system	Discharge rate	Efficiency and power output
[17, 18]	Solar air heater	Solar irradiance, temp. and flow rate	Heat gain and efficiency
[This Paper]	Directly connected Helical rotor WPS system	Irradiance, voltage, current, and head	Efficiency, average flow, motor speed, and power output

1.2 Water Pumping System

PV based WPS utilizes sun energy and transforms it into mechanical energy which is converted into hydraulic energy by the pump. Pumps are mainly classified into submersible, floating, and surface water pumps. In the studied system, helical rotor-based submersible motor is used. A submersible motor extracts the water from deep well and stores it in the storage tank for later use.

The energy supplied to motor depends upon the power extracted from PV panels. MPPT is integrated with the system to draw the maximum available power from PV array. MPPT checks the irradiance, ambient temperature, PV array orientation, and time of the day. Based on the measurement, it regulates the operating voltage of the PV array to extract the maximum available power for the system. The current–voltage relation of a single PV cell can be given as in Eq. (1.1).

$$I = I_{\mathrm{L}} - I_{\mathrm{o}}\left\{\exp\left(\frac{q}{akT_{\mathrm{c}}}(V + IR_{\mathrm{s}})\right) - 1\right\} - \frac{V + IR_{\mathrm{s}}}{R_{\mathrm{sh}}} \tag{1.1}$$

where I_{L}—current generated due to light, I_{o} is reverse saturation current, a is ideality factor which vary in the range of 1–5, k represents Boltzmann constant, and T_{c} is cell temperature. R_{sh} and R_{s} indicate parallel and series resistance values, respectively.

The system taken for the analysis is shown in Fig. 1.1. Solarex panels (MSX-60) are utilized with two panels connected in series. Proposed PV array is capable of delivering a total power of 120 W at an irradiance of 1000 W/m^2 and ambient temperature of 25°. The parameters of the PV module is given in Table 1.2.

Hydraulic power requires to pump a certain amount of water, which depends on the head of the system and the water flow rate. The above relation is expressed in Eq. (1.2) [19].

$$P_{\mathrm{h}} = Q * H * g * \rho \tag{1.2}$$

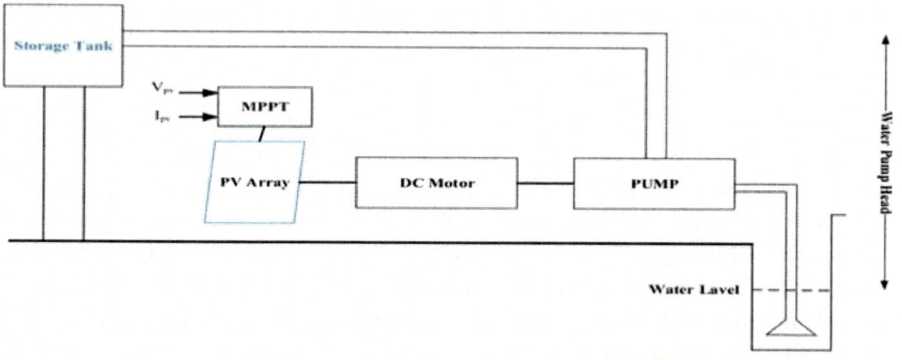

Fig. 1.1 Simulated PV based water pumping system

Table 1.2 Electrical characteristics of solar panel MSX-60

Maximum power	60 W
Voltage at P_{max} (V_{mp})	17.1 V
Current at P_{max} (I_{mp})	3.5 A
Short-circuit current (I_{sc})	3.8 Amps
Open-circuit voltage (V_{oc})	21.1 V
Temperature coefficient of I_{sc}	(0.065 ± 0.015)%/°C
Temperature coefficient of V_{oc}	$-(80 \pm 10)$mV/°C
Maximum series fuse rating	20 Amps

where Q is the flow rate (l/sec), H represents head of the system, g is gravitational acceleration, and ρ represents water density. The flow rate Q can be calculated using Eq. (1.3).

$$Q = V_{gd} * \omega \tag{1.3}$$

$$\eta_v = \frac{Q - Q_L}{Q} \tag{1.4}$$

$$\eta_m = \frac{P_{th}}{P_{th} + P_{fr}} \tag{1.5}$$

$$\eta_{system} = \eta_m * \eta_v \tag{1.6}$$

The system's overall efficiency is multiplication of volumetric and mechanical efficiency, which is expressed in Eqs. (1.4)–(1.6). V_{gd} is volumetric displacement of the pump (rad/sec) and ω is motor speed, where Q_L is leakage rate (l/s), P_{th} is calculated required power, and P_{fr} represents losses due to friction.

1.3 Development of ANN Network

ANN is a robust technique that can be applied in wide area of applications. For the analysis, a feedforward backpropagation ANN structure is designed. This is well-known and developed method which can form a good relationship for complex, nonlinear systems. A three-layer network is shown in Fig. 1.2, which consists of an input layer, a hidden layer, and an output layer. The input of ANN network measures the PV insolation value, operating voltage, operating current, and water pumping head. Output layer having four neurons represents overall system efficiency, average flow rate, power output, and rotor speed.

In the shown ANN structure, blue-colored circles are in the input layer; green-colored circles are present in the output layer, and the middle layer called the hidden layer encompasses all the yellow circles. Each circle represents neurons which are

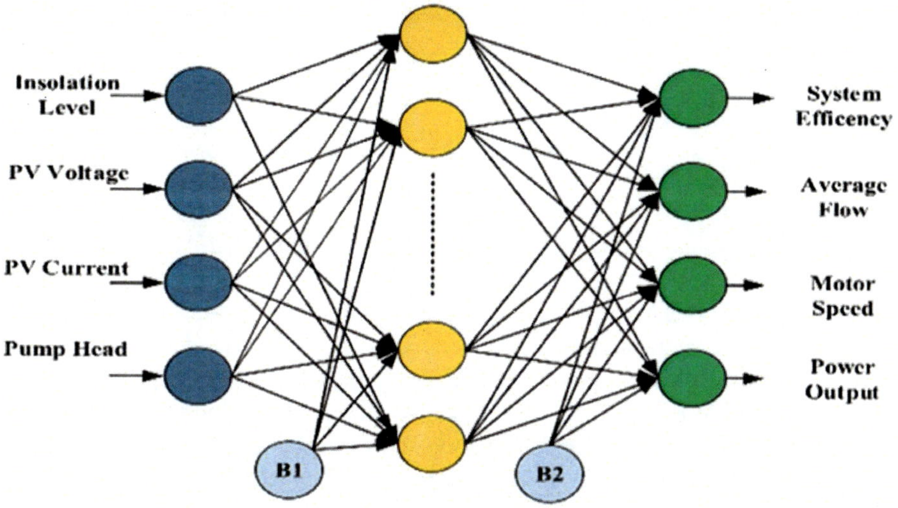

Fig. 1.2 Developed ANN for system performance analysis

connected with its proceedings neurons with a value called synaptic weights. The gray color circle represents bias value at the respective layer. Output of the ANN system can be expressed as in Eq. (1.7), where i represents ith output and w_{ik} is the synaptic weights between ith and kth neurons.

$$y_i = \sum_{k=1}^{n} (\omega_{ik}) + b_i \tag{1.7}$$

Levenberg–Marquardt (LM) algorithm is used to train the network and to optimize the weights between the neurons. For ANN training, experimental field data [19] is considered. Out of 120 available field data, 65 data were selected to train the network. The number of neurons and layers was evaluated with repeated training and remarking the error obtained through each specified network. The developed system found to be efficient, having one layer of hidden layer with 10 neurons. Tan-sigmoid transfer function is applied at hidden layer and linear transfer function at output layer, and the developed network was appropriate for the performance analysis. On the final step, after selecting the most appropriate network, 50 sample data that were not identical with the training data were tested.

1.4 Results and Discussion

To analyze the PV based WPS with helical rotor, the developed ANN model is used. It is of interest to see the effects on efficiency of the system, average flow, power output, and motor speed variation corresponding to change in the head of

Table 1.3 Performance of ANN developed for analyzing purposes

	Mean squared error	R^2
Training	2.31569e−5	0.99993
Validation	1.06158e−4	0.99999
Testing	1.41510e−4	0.99989

the pumping system. Table 1.3 represents the performance of ANN developed for analyzing purposes.

R^2 ranges between 0 and 1 represent the relation between the input and output, higher the value, more substantial the connection is. We obtained the approximately = 1 for the trained ANN network, which signifies the solid bound between four inputs and four outputs. Table 1.3 illustrates the mean squared error and R^2 value obtained in the respective stage during the ANN development.

For system's performance analysis, water pumping head (base condition of 23 m) is varied from 20 to 150%, and correspondingly percentage change in efficiency, motor speed, average flow, and power output is measured through developed ANN. Insolation level, operating voltage, and current of the PV are kept constant at 960 W/m^2, 29 V, and 3.12 A, which corresponds to maximum power point at given irradiance.

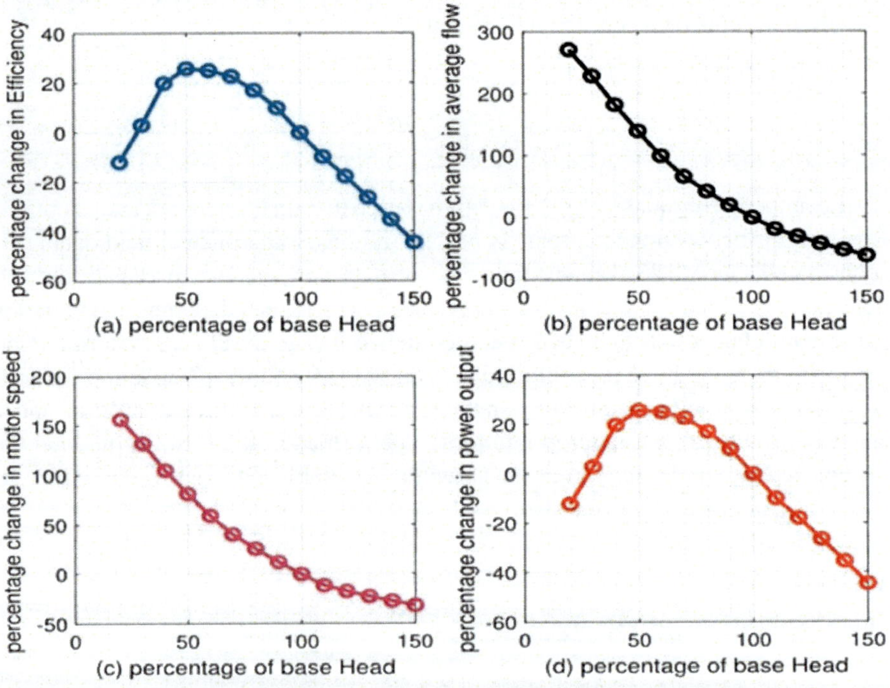

Fig. 1.3 Effects on **a** efficiency, **b** average flow, **c** motor speed, and **d** power output

Figure 1.3 represents the percentage change in parameter compared to its base value with pump head variation. As (from Fig. 1.3) observation is made that when head is kept at 50% of its base value, the efficiency, average flow, motor speed, and power output increases by 20%, 140%, 85%, and 25%, respectively, from its base value. The similar percentage variation is incorporated, and the changes are plotted in (from Fig. 1.3). Points that can be inferred from the observations are

- The efficiency and power output of the system are interrelated. When the head is below 40% of the base value, the efficiency and power output are significantly low compared to the base condition.
- Efficiency and power output can be improved by slightly decreasing the pumping head (approximately 50% of base value) if possible, which enhances the overall performance of the system.
- Efficiency and so is the power output first increase and then decrease with increasing the head of the pumping system.
- It is noted that the average flow of water depends upon motor speed. The higher the rate, the more will be the flow of water.
- An increase in head leads to a decrease in motor speed and correspondingly decreased average flow, which is evident as it requires higher power than the rated condition.

1.5 Conclusion

Effects on overall efficiency, average flow, power output, and motor speed are observed through variation in the head of the WPS. The studied work focuses on the development of a tool which can be utilized to get the system performance with a short time and better accuracy. It is observed that power output, speed, average flow, and efficiency all are interrelated with pump head of the WPS. If the head is chosen below 30% of its base value, it impacts the efficiency and performance to get worst as compared to base working condition. So choosing the proper head is of great importance. Also, the ANN tool suits to be the decent tool for the proposed work which can analyze the system accurately and determine the performance parameter of the WPS. The ANN tool and its other structure can be explored to solve other power system complexities.

References

1. R. Posorski, Photovoltaic water pumps, an attractive tool for rural drinking water supply. Sol. Energy **58**(4–6), 155–163 (1996)
2. A.M. Zaki, M.N. Eskander, Matching of photovolatic motor-pump systems for maximum efficiency operation. Renew. Energy **7**(3), 279–288 (1996)
3. N. Chandrasekaran, K. Thyagarajah, Modeling and performance study of single phase induction motor in PV fed pumping system using MATLAB. Int. J. Electr. Eng. **5**(3), 305–316 (2012)

4. S. Chandra, P. Gaur, P. Diwaker, Radial basis function neural network based maximum power point tracking for photovoltaic brushless DC motor connected water pumping system. Comput. Electr. Eng. **86**, 106730 (2020)
5. N. Chandrasekaran, B. Ganeshprabu, K. Thyagarajah, Comparative study of photovoltaic pumping system using a DC Motor and PMDC motor, in *Proceedings of IEEE International Conference on Advances in Engineering, Science and Management. ICAESM-2012* (2012), pp.129–132
6. B.D. Vick, R.N. Clark, Water pumping performance of a solar-PV helical pump, in *Proceedings of ISES 2005 Solar World Congress: Solar Energy—Bringing Water to the World* (2005), pp. 6–12
7. W. Lawrance, B. Wichert, D. Langridg, Simulation and performance of a photovoltaic pumping system, in *Proceedings of 1995 International Conference on Power Electronics and Drive Systems* (1995), pp. 513–518
8. G. Biji, Modelling and simulation of PV based pumping system for maximum efficiency, in *2012 International Conference on Power, Signals, Controls and Computation* (IEEE, India, 2012)
9. B. Li, C. Delpha, D. Diallo, A. Migan-Dubois, Application of artificial neural networks to photovoltaic fault detection and diagnosis: a review. Renew. Sustain. Energy Rev. 110512 (2020)
10. F. Fachini, B.I.L. Fuly, A Comparison of machine learning regression models for critical bus voltage and load mapping with regards to max reactive power in PV buses. Electric. Power Syst. Res. **191**, 106883 (2021)
11. S. Srinivasan, R. Tiwari, M. Krishnamoorthy, M.P. Lalitha, K.K. Raj, Neural network based MPPT control with reconfigured quadratic boost converter for fuel cell application. Int. J. Hydrogen Energy (In press) (2020)
12. A.R. Pazikadin, D. Rifai, K. Ali, M.Z. Malik, A.N. Abdalla, M.A. Faraj, Solar irradiance measurement instrumentation and power solar generation forecasting based on Artificial Neural Networks (ANN): A review of five years research trend. Sci. Total Environ. **715**, 136848 (2020)
13. A. Mokeddem, A. Midoun, D. Kadri, H. Said, A.R. Iftikhar, Performance of adirectly-coupled PV water pumping system. Energy Conv. Manag. **52**(10), 3089–3095 (2011)
14. O. Atlam, M. Kolhe, Performance evaluation of directly photovoltaic powered DCPM (direct current permanent magnet) motor—propeller thrust system. Energy **57**, 692–698 (2013)
15. M. Jafar, A model for small-scale photovoltaic solar water pumping. Renew. Energy **19**(1–2), 85–90 (2000)
16. R. Mandal, R. Naskar, A study of solar photovoltaic application in irrigation system and its performance analysis in laboratory scale. Int. J. Adv. Altern. Energy Environ. Ecol. **1**(1), 1–14 (2012)
17. H.K. Ghritlahre, K. Prasad, Prediction of thermal performance of unidirectional flow porous bed solar air heater with optimal training function using artificial neural network. Energy Proc. **109**, 369–376 (2017)
18. H.K. Ghritlahre, R.K. Prasad, Application of ANN technique to predict the performance of solar collector systems—a review. Renew. Sustain. Energy Rev. **84**, 75–88 (2018)
19. R.E. Katan, V.G. Agelidis, C.V. Nayar, Performance analysis of a solar water pumping system, in *Proceedings of International Conference on Power Electronics, Drives and Energy Systems for Industrial Growth, NewDelhi, India* (1996), pp. 81–87

Chapter 2
A Comparative Analysis of Neural Network-Based Models for Forecasting of Solar Irradiation with Different Learning Algorithms

Anuj Gupta, Kapil Gupta, and Sumit Saroha

Abstract In the smooth operation of standalone and grid connected solar power plants, solar forecasting plays an important role and develops bilateral contract agreement between suppliers and customers. An efficient forecasting technique will provide great help to grid operators as it will balance the electricity demand and generation. In this paper, two neural network models: cascade forward back propagation neural network (CFNN) and Elman back propagation neural network (Elman) are established with four different algorithms. The performance of model's are being judged on the basis of mean absolute percentage error (MAPE), root mean square error (RMSE), and correlation coefficient (R^2).The meteorological data of two years from Delhi location are used for training, and one year data on the seasonal basis are used for testing the network. Elman back propagation neural network model perform better as a comparison to cascade forward back propagation neural network model. The average value of MAPE, RMSE, and R^2 using the Elman back propagation neural network with LM algorithm is 11.38%, 22.65%, and 88.11%, respectively. The MAPE's lowest value informs us about the forecasting model's efficiency. The proposed neural network models have the least MAPE and RMSE used to forecast global solar irradiation.

Keywords Artificial neural network · Statistical error · Back propagation algorithm · Solar irradiation

2.1 Introduction

Due to the increased demand of electricity, the use of solar energy also increases. The received solar energy is directly used to generate the electricity using solar PV panel [1]. However, the weather variability's and climatic conditions proportionality

A. Gupta (✉) · K. Gupta
Maharishi Markandeshwar (Deemed To Be University), Mullana-Ambala, India

S. Saroha
Guru Jambheshwar University of Science and Technology, Hisar, India

A. Khosla and M. Aggarwal (eds.), *Smart Structures in Energy Infrastructure*,
Studies in Infrastructure and Control,
https://doi.org/10.1007/978-981-16-4744-4_2

affect performance of solar PV panel's results the random electricity generation. The variable and uncertain nature of solar PV panels is also a danger for grid instability, balancing, and operation. Forecasting of solar energy in advance for a particular location not only gives the electricity generation from solar PV panel in advance but also improves the grid stability [2]. Various models have been developed in the literature to forecast solar irradiation at various origin along with different preprocessing technique which have also been applied to the model to improve the forecasting accuracy [3]. Mohammed Bou-Rabee et al. estimated the solar radiation by gradient descent method and Levenberg–Marquardt (LM) back propagation algorithm. The accuracy of this model was determined by the MAPE which was 86.3% for the gradient descent method and 85.6% for LM [4].

The six different ANN models with different combination of input were created by Ahmat Koca et al. to predict the solar irradiation. The six models have inputs from latitude, longitude, altitude, average cloudiness, and average humidity. The minimum RMSE for the designed model was 0.0358, whereas R^2 was 0.9974 [5]. M. A. Behrang et al. developed a model to predict the DGSR using ANN based algorithm. The six models have been designed to estimate the DGSR with different combination of meteorological input variables [6]. Moreover, proper selection of training algorithm and parameters of the models is still a big challenge to design a forecasting model with improved accuracy [7]. Therefore, this paper aims to access the performance of two ANN based model: CFNN, Elman on the basis of different training algorithm: Levenberg–Marquardt (LM), scaled conjugate gradient (SCG), gradient descent (GD), and resilient back propagation (RP). The study also observes the effect of meteorological variables on the forecasting performance of the model. Different meteorological variables: temperature, relative humidity, pressure, solar zenith angle, wind speed, wind direction, and precipitation water, are used in the study as an input to the models, while solar GHI is used as the output of the model. The performance of the model is evaluated using MAPE, RMSE, and R^2.

The remaining paper is organized as follows. Section 2.2 describes the artificial neural network model for solar irradiation forecasting. Section 2.5 explains the methodology used in the paper. The result and discussion are presented in Sect. 2.6. Section 2.7 describes the selection of best training algorithm and the conclusion is present in Sect. 2.8

2.2 ANN Model for Solar Irradiation Forecasting

The working of ANN is almost comparable to the human brain which takes the decision based on the biological neurons [8]. Different types of parallel processing, pattern recognition analysis are performed by the neurons in the human brain. The same phenomena can be applied to solve nonlinear mathematical problem in image processing, forecasting, etc. [9]. This technique-based model has to train repeatedly to obtain the best value to map the input and output. In this study, two ANN models are used to forecast solar irradiation for Delhi location. Seven meteorological parameters

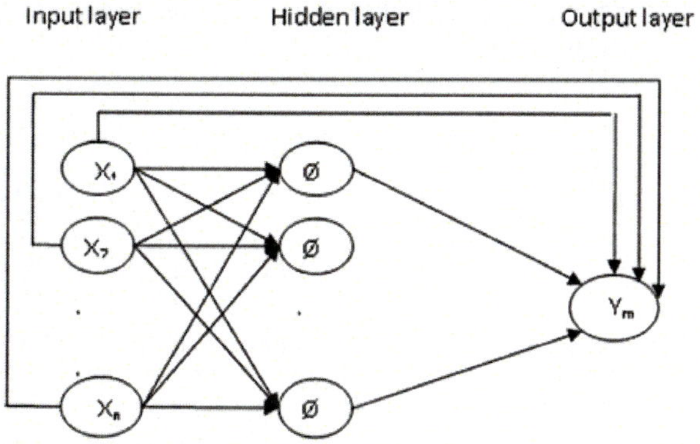

Fig. 2.1 Architecture of cascade forward neural network [10]

are used as an input to the models, and four different algorithms are used to check the performance of the model. The number of hidden neuron varies from 1 to 30.

2.3 Cascade Forward Back Propagation Neural Network

It is also known as "self-organizing" networks. It is a type of neural network which is almost same as feed forward network but it is having a link from the past and input layer to following layers. The basic architecture of cascade forward neural network is shown in Fig. 2.1.

Cascade forward neural network consist of three layers, the output layer is directly linked with the input layer along with hidden layer, and the input layer distributes the input value to the hidden and output layer. In addition, a bias value of 1.0 is fed to the output and hidden layer. In the hidden layer, each input value is multiplied by a weight, combined them, and transferred into a transfer function [10]. The output of the hidden layer is fed to an output layer that receives value from the input neurons, including bias and value from the neurons of the hidden layer [11].

2.4 Elman Back Propagation Neural Network

It is developed by Elman in 1990. It consist of four layer (i) input layer, (ii) hidden layer, (iii) output layer, and (iv) context layer. The input layer uses neurons for transmitting the signal, while the neurons of the output layer play a linear weighted role and use a purelin activation function. The hidden layer transfer mode is still a nonlinear method which is the sigmoid function. The input layers connect to the output of hidden layer through the context layer's delay and memory. It is also named as a recurrent neural network and differs from a feed forward neural network because it contains a context layer. Elman network is also used for solving prediction problem of discrete time sequence, but it also faces several difficulties like as low

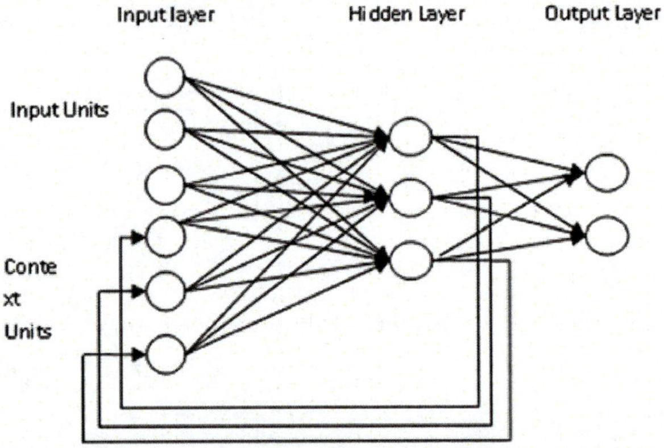

Fig. 2.2 Structure of Elman neural network [12]

Table 2.1 Geographical location of Delhi

S. no.	Place	Latitude/Longitude	Altitude (m)
1	Delhi	28°38′/77°13′	216

generalization problem and higher speed of convergence [12]. The configuration of Elman neural network is shown in Fig. 2.2.

2.5 Methodology

2.5.1 Data Collection

In India, due to the high cost of measuring devices, it is not possible to measure solar radiation data. Hourly data for pressure, temperature, relative humidity, precipitation water, solar zenith angle, wind speed, and wind direction of Delhi location are collected for three years (2012–14) from National Solar Radiation Database (NSRDB). After collecting the meteorological variables, the data are divided into training and testing set. Two year data (2012–13) are used for training, and one year data (2014) are divided into seasonal basis for testing. Geographical location of Delhi station [13] is shown in Table 2.1.

2.5.2 ANN Implementations

In this study, two neural network models CFNN and ELMNN are implemented in MATLAB version 2019. Seven meteorological parameters temperature, relative humidity, pressure, solar zenith angle, wind speed, wind direction, and precipitation water are used as an input, and only one parameter global horizontal irradiation is forecasted as output. Both models are trained several times to get the best value. The value of hidden neuron varies from 1 to 30.The most common algorithms Levenberg–Marquardt, scaled conjugate gradient, gradient descent, and resilient back propagation are used for training a neural network model [14]. The statistical metrics MAPE, RMSE, and R^2 are used to check the performance of forecasting model.

2.6 Results and Discussion

The main aim of this paper is to develop an accurate forecasting model for the prediction of solar irradiation. The performances of neural network models are being judged on the basis of MAPE, RMSE, and R^2 which is expressed by the equation [12, 15].

$$\text{MAPE} = \left(\frac{1}{n}\sum_{i=1}^{n}\frac{\left|\text{GSR}_{i\,(\text{predicted})} - \text{GSR}_{i\,(\text{actual})}\right|}{\text{GSR}_{i\,(\text{actual})}}\right) \times 100 \tag{2.1}$$

$$\text{RMSE} = \sqrt{\frac{1}{n}\sum_{i=1}^{n}\left(\text{GSR}_{i\,(\text{predicted})} - GSR_{i\,(\text{actual})}\right)^2} \tag{2.2}$$

$$R^2 = \left(1 - \left(\frac{\sum_{i=1}^{n}\left|\text{GSR}_{i\,(\text{predicted})} - \text{GSR}_{i\,(\text{actual})}\right|^2}{\sum_{i=1}^{n}\text{GSR}_{i\,(\text{actual})}}\right)\right) \times 100 \tag{2.3}$$

The MAPE represent the uniform prediction error in percentage, while RMSE indicates the divergence between predicted and observed value, and R correlation coefficient represents the relation between measured and forecast value. The lowest value of MAPE and RMSE indicates more accurate in estimation of solar irradiation. However, a lowest value of MAPE, RMSE, and high value of correlation coefficient (R) indicates the best proposed model for forecasting. The most efficient algorithm for forecasting of global horizontal irradiation was obtained according to the results given in Tables 2.2 and 2.3. From the result, it is seen that Elman neural network performs better as comparison to cascade forward neural network in terms of MAPE, RMSE, and R^2. Both models trained using four algorithms, and at the time of analysis, LM algorithm performs better as a comparison to other algorithms. It is clear that from the analysis of result, Elman neural network LM algorithm suitable for predicting the global horizontal irradiation.

Table 2.2 Error metrics for CFNN model

Name of the algorithm	Seasons	MAPE (%)	RMSE (W/m^2)	R^2 (%)
Levenberg–Marquardt (LM)	Winter	14.0314	29.0399	84.74
	Spring	15.9706	31.0093	83.26
	Summer	8.9656	20.8674	91.05
	Monsoon	18.8656	37.7826	79.06
	Autumn	8.2781	20.1538	93.02
Average value		13.22226	27.7706	86.226
Scaled conjugate gradient (SCG)	Winter	16.4909	31.1156	82.45
	Spring	17.0314	32.9933	83.55
	Summer	9.5910	22.0192	93.06
	Monsoon	19.3855	38.9102	77.09
	Autumn	9.9129	23.8765	92.55
Average value		14.48234	29.7829	85.74
Resilient back propagation (RP)	Winter	17.4004	33.7928	80.55
	Spring	18.4558	34.0416	79.74
	Summer	10.3038	23.1033	90.49
	Monsoon	21.7137	41.9102	76.33
	Autumn	10.8964	24.7137	88.16
Average value		15.75402	31.51232	83.054
Gradient descent (GD)	Winter	24.4550	38.4844	76.62
	Spring	27.9545	47.9272	71.13
	Summer	14.7113	27.5852	83.32
	Monsoon	29.8964	49.8322	69.64
	Autumn	15.7656	29.0416	81.08
Average value		22.55656	38.57412	76.358

2.7 Selection of the Best Training Algorithm

A comparative analysis between two neural network-based models indicates that both model suitable for predicting solar irradiation when Levenberg–Marquardt is used for training and testing. However, meteorological parameters used for the current study are site specific. Therefore, it is important to find the best neural network model to predict solar irradiation with minimum error. The best performance of Elman neural network with LM algorithm is given in Table 2.4.

The performance plot of predicted GHI and measured GHI for all seasons of Elman-LM model are shown in Figs. 2.3, 2.4, 2.5, 2.6, and 2.7. It can be seen from the figure that the ANN forecasted values for almost all dataset are very similar to the observed global solar radiation values.

Table 2.3 Error metrics for Elman model

Name of the algorithm	Seasons	MAPE (%)	RMSE (W/m^2)	R^2 (%)
Levenberg–Marquardt (LM)	Winter	14.6590	23.7499	84.95
	Spring	13.2229	21.2619	85.54
	Summer	7.8200	19.3315	92.59
	Monsoon	15.0825	31.4690	79.95
	Autumn	6.1243	17.4670	97.53
Average value		11.38174	22.65586	88.112
Scaled conjugate gradient (SCG)	Winter	16.0038	31.6076	81.08
	Spring	15.5676	29.6201	86.59
	Summer	9.7891	24.9874	91.85
	Monsoon	17.6351	38.4383	78.45
	Autumn	7.1714	20.9011	95.04
Average value		13.2334	29.1109	86.602
Resilient back propagation (RP)	Winter	18.7174	34.4714	79.88
	Spring	15.3013	31.5061	81.44
	Summer	9.5629	25.6680	92.05
	Monsoon	18.4159	40.2619	80.05
	Autumn	8.7408	19.7278	94.66
Average value		14.14766	30.32704	85.616
Gradient descent (GD)	Winter	24.3400	39.1412	76.33
	Spring	26.8595	45.1915	72.74
	Summer	13.0399	27.1852	84.04
	Monsoon	28.3454	47.1322	68.14
	Autumn	14.4120	29.1416	81.16
Average value		21.39936	37.55834	76.482

Table 2.4 Selection of best training algorithm

LM algorithm			
Model	MAPE (%)	RMSE (W/m^2)	R^2 (%)
ELMNN-LM	**11.38174**	**22.65586**	**88.112**
CFNN-LM	13.22226	27.7706	86.226

2.8 Conclusion

In this paper, two neural network-based models were implemented for the forecasting of solar irradiation. Both models were trained and tested using four back propagation algorithm: Levenberg–Marquardt, scaled conjugate gradient, resilient back propagation, and gradient descent. Two year data were used for training, and one year data on seasonal basis were used for testing the network. The evaluations

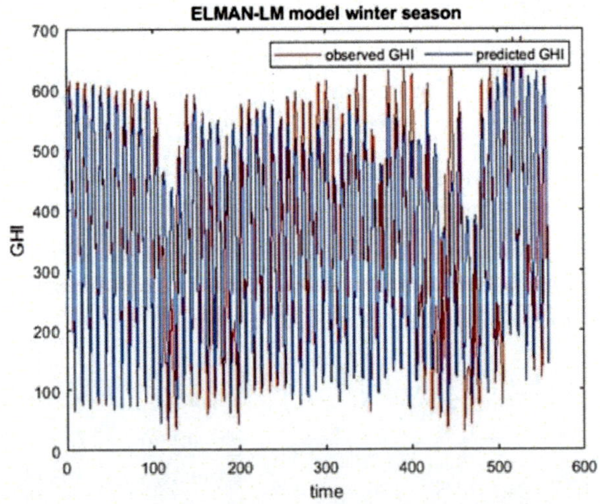

Fig. 2.3 Elman-LM model winter season

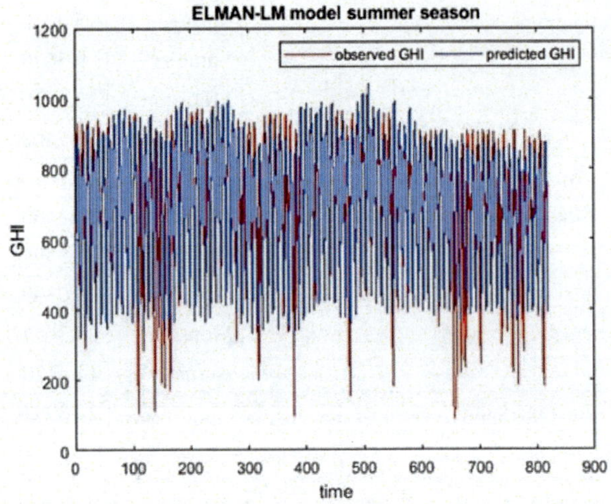

Fig. 2.4 Elman-LM model summer season

of best model and LM algorithm are performed on the basis of minimum value of mean absolute percentage error, root mean square error, and maximum linear correlation coefficient. From the result analysis, it is seen that both models are suitable for forecasting purpose, but when we make a comparative analysis, Elman neural network model performs better with LM algorithm. Finally, the result analysis indicates that Elman neural network is best for predicting the solar irradiation in India where meteorological station is not available.

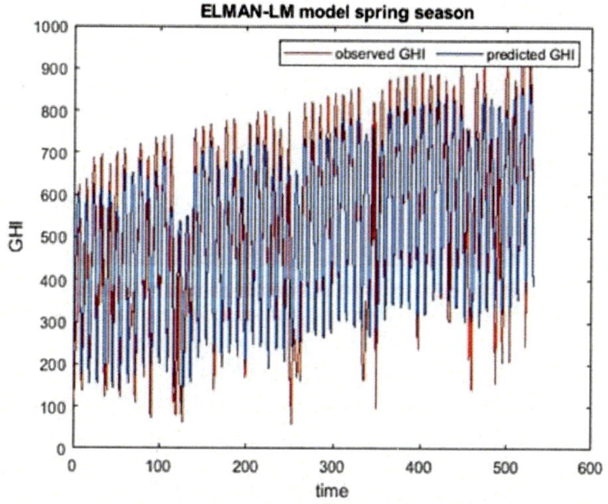

Fig. 2.5 Elman-LM model spring season

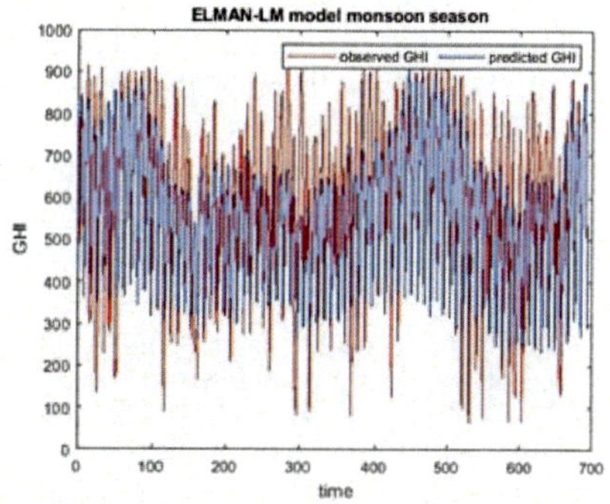

Fig. 2.6 Elman-LM model monsoon season

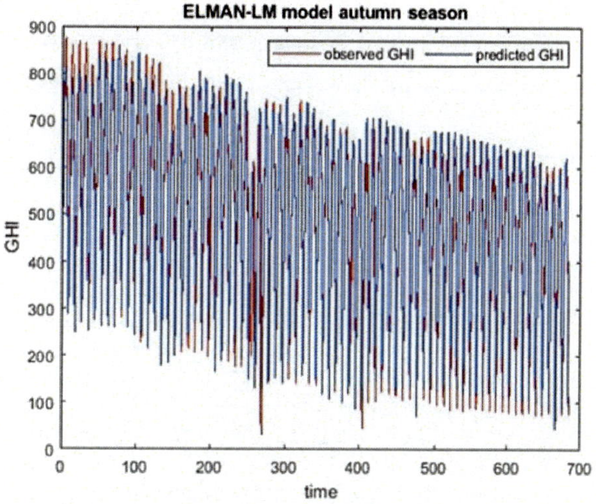

Fig. 2.7 Elman-LM model autumn season

References

1. P. Singla, M. Duhan, S. Saroha, A comprehensive review and analysis of solar forecasting techniques. Front. Energy. https://doi.org/10.1007/s11708-021-0722-7
2. A. Quammi, D. Zejli, H. Dagdougui, R. Benchrifa, Artificial neural network analysis of Moroccan solar potential. Renew. Sustain. Energy Rev. **16**, 4876–4889 (2012)
3. M. Mohandes, M. Rehman, S. Rehman, T.O. Halawani, Estimation of global solar radiation using artificial neural networks. Renew. Energy **14**, 179–184 (1998)
4. M. Bou-Rabee, S.A. Sulaiman, M.S. Saleh, S. Marafi, Using artificial neural network to estimate solar radiation in Kuwait. Renew. Sustain. Energy Rev. **72**, 434–438 (2017)
5. A. Koca, H.F. Oztop, Y. Varol, G.O. Koca, Estimation of solar radiation using artificial neural networks with different input parameters for Mediterranean region of Anatolia in Turkey. Expert Syst. Appl **38**(7), 8756–8762 (2011)
6. M.A. Behrang, E. Assareh, A. Ghanbarzadeh, A.R. Noghrehabadi, The potential of different ANN technique in daily global solar radiation modeling based on data. Sol. Energy. **84**(8), 1480–1486 (2010)
7. S. Kumar, T. Kaur, Developments of ANN based model for solar potential assessment using various meteorological parameters. Energy Procedia **90**, 587–592 (2016)
8. C.D. Lewis, International and Business Forecasting Methods (Butter-worths, Lodon, 1982)
9. N. Kumar, U.K. Sinha, S.P. Sharma, Y.K.N. Nayak, Prediction of daily global solar radiation using neural network with improved gain factor and RBF Network. Int. J. Renew. Energy Res. **7**, 1235–1244 (2017)
10. B. Warsito, R. Santoso, S. Suparti, H. Yasin, Cascade forward neural network for time series prediction. IOP Conf. Series J. Phys. Conf. Series 1025 (2018)
11. B. Siveaneasan, C. Yu, K.P. Goh, Solar forecasting using ANN with fuzzy logic pre processing. Energy Procedia**143**, 727–732 (2017)
12. I. Khan, H. Zhu, J. Yao, D. Khan, Photovoltaic power forecasting based on elman neural network software engineering method, in *8th IEEE International Conference on Software Engineering and Service Science (ICSESS)* (2017)
13. http://delhitourism.gov.in/delhitourism/aboutus/seasons_of_delhi.jsp
14. N. Premalatha, A. ValanArasu, Prediction of solar radiation for solar systems by Using ANN models with different back propagation algorithm. J. Appl. Res. Technol. **14**(3), 206–214 (2016)
15. S. Sobri, S. koohi-kamali, N.A. Rahim, Solar photovoltaic generation forecasting methods: A review. Energy Conver. Manag. **156**, 459–497 (2018)

Chapter 3
A Review of Load Frequency Control of Hybrid Power System

Amit Atri and Anita Khosla

Abstract Need of load frequency control (LFC) comes due to the continuous variations in the load correspondingly varying the system frequency. Initially, conventional PID controllers were used, and then, the parameters of these controllers were optimized using advanced optimization techniques and the use of artificial intelligence become prominent. With the involvement of renewable energy-based power generation, the continuous supply of the input becomes an issue, and the use of storage devices in the power system becomes important. This review paper presents different types of controllers used for load frequency control, different optimization techniques used for optimizing the parameters of the controllers, and the use of different types of storage systems used in hybrid power systems.

Keywords Automatic generation control (AGC) · Load frequency control (LFC) · Proportional integral derivative control (PID) · Artificial intelligence (AI) · Energy storage system (ESS)

3.1 Introduction

Electric energy is preferred over other sources of energy due to its numerous advantages such as neither it has any byproducts nor it causes any kind of pollution. Electric energy can very easily be converted into another from of energy, it is economically efficient, can easily transmitted over long distances, and easy to control and monitor. But one of the major disadvantages of electric power is the limited storage. So, it is a constraint of power system's reliable operation that every time the generated power must equal the load. There are many active and passive elements interconnected in a power system. Connected load in a power system varies continuously. This causes fluctuations in the active power and system frequency. For maintaining

A. Atri (✉) · A. Khosla
Manav Rachna International Institute of Research and Studies, Faridabad, India

A. Khosla
e-mail: anitakhosla.fet@mriu.edu.in

A. Khosla and M. Aggarwal (eds.), *Smart Structures in Energy Infrastructure*,
Studies in Infrastructure and Control,
https://doi.org/10.1007/978-981-16-4744-4_3

the power system quality, the active power needs to be balanced, and so the system frequency remains within acceptable limits. A total of 50 Hz is the standard frequency in India, target frequency range is 49–50 Hz, and the acceptable frequency range is 48.5–51.5 Hz.

Owing to the various advantages that renewable energy offers like low maintenance, economical operation, environment friendly and long lasting, it is gaining more space in power generation sector. It makes the power system more distributed and uncertain as well due to the weather and environmental conditions.

Due to the various advantages and disadvantages of renewable sources, they are combined with conventional energy sources for power generation. Such power systems consisting of both renewable and conventional sources based power generation are known as hybrid power system. The renewable source used in hybrid power system are mostly solar photovoltaic or a wind turbine [1]. Biomass-based power generation can also be used especially in countries like India where there is a huge availability of biomass. But supply of both solar and wind energy is inconsistent which makes the operation of an isolated renewable energy-based power system unreliable. To make a renewable power plant reliable, it is connected with a grid or it is operated alongwith another power generation or compensation unit (hybrid power system).

The hybrid power system deals with the frequency stabilization not only due to load change but also change in power generated. The geographical diversity in case of the distributed generation causes some issues like time delay and packet dropouts. In case of a solar panel, the area of the panel under solar exposure directly affects the electrical output power. A generation control system is needed which can maximize the output [2]. Another approach is to provide compensation so as to minimize the variations in the output of the renewable source unit [3]. Another approach for a wind turbine is the use of a control system which tracks an absolute power command which maximizes the power captured if the wind velocity reduces and also maintains a specified power reserve proportional to available power in the wind [4].

Automatic generation control (AGC) shown in Fig. 3.1 becomes most important control system which maintains the system frequency at a nominal value (within acceptable limits) and maintains the correct amount of power interchange between different power systems meanwhile maintaining each units generation at the most economical value. When load is shared by both the renewable and conventional source, then generation control becomes more complicated. A control system (for controlling the load frequency) is required, which detects the changes in the load (corresponding to the frequency change), adjusts the generation, and thereby stabilizes the system frequency. Such a system is shown in Fig. 3.1.

So as to maintain the system frequency within stable limits, various controllers are implemented. The following equation describes the performance of a conventional speed governing system

$$\Delta Y_{\mathrm{E}}(s) = \left[\Delta P_{\mathrm{c}}(s) - \frac{1}{R}\Delta F(s)\right] \times \left(\frac{K_{\mathrm{sg}}}{1 + T_{\mathrm{sg}}s}\right) \tag{3.1}$$

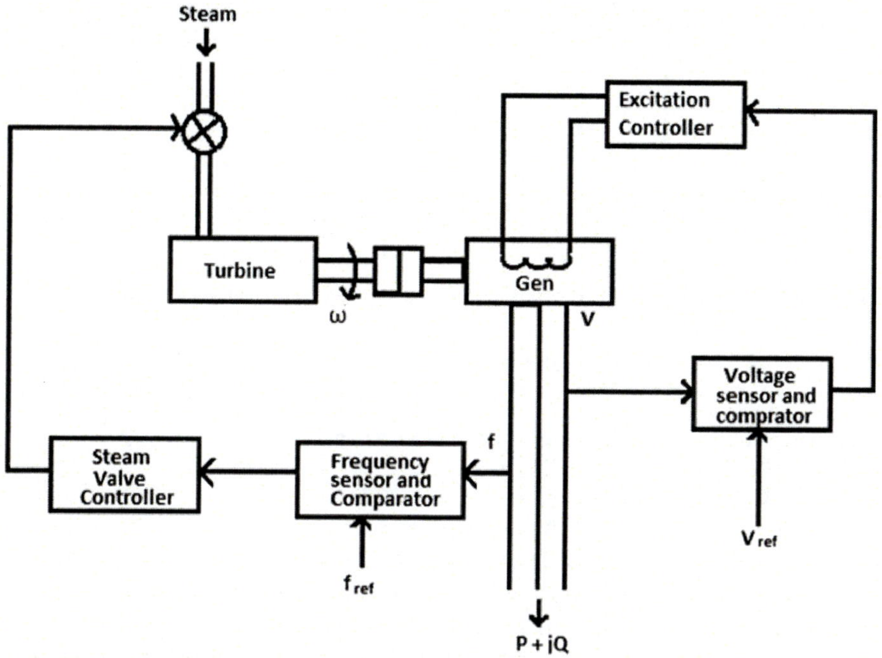

Fig. 3.1 Load frequency controller and excitation voltage regulators of a generator

ΔY_E = change in steam valve setting
ΔP_C = commanded increase in power
ΔF = change in frequency
R = speed regulation of the governor
K_{sg} = gain of speed governor
T_{sg} = time constant of the speed governor.

The change in the steam valve setting ΔY_E causes a change in the turbine power ΔP_t which depends on the turbine gain K_t and time constant T_t as:

$$\Delta P_t(s) = \Delta Y_E(s)\left(\frac{K_t}{1 + T_t s}\right) \tag{3.2}$$

If P_G is the power generated by the generator, then $\Delta P_G = \Delta P_t$. P_D is the power demand, and then any change in frequency is

$$\Delta F(s) = [\Delta P_G(s) - \Delta P_D(s)] \times \left(\frac{K_{ps}}{1 + T_{ps} s}\right) \tag{3.3}$$

T_{ps} = power system time constant
K_{ps} = power system gain.

The speed regulation R is so adjusted that changes in frequency are small, and so there is a linear relationship between frequency and load as shown in Fig. 3.2.

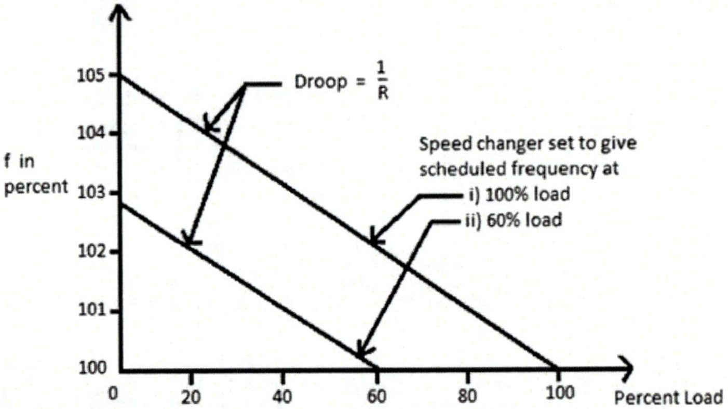

Fig. 3.2 Load frequency characteristic of a speed governing system

Proportional integral derivative controllers are used widely for load frequency control. They are having various advantages such as low cost, simple structure, provide good stability, rapid response, and are stable. Although we need to tune its parameters [5], various optimization techniques are used for that like moth flame optimization, elephant herding optimization, etc. Also other artificial intelligence-based controllers are used for load frequency control [6].

3.2 Load Frequency Control of Wind-Diesel Hybrid Power System

Considering a hybrid power system consisting of a wind turbine generator (WTG) combined with a diesel engine generator (DEG) shown in Fig. 3.3, combining

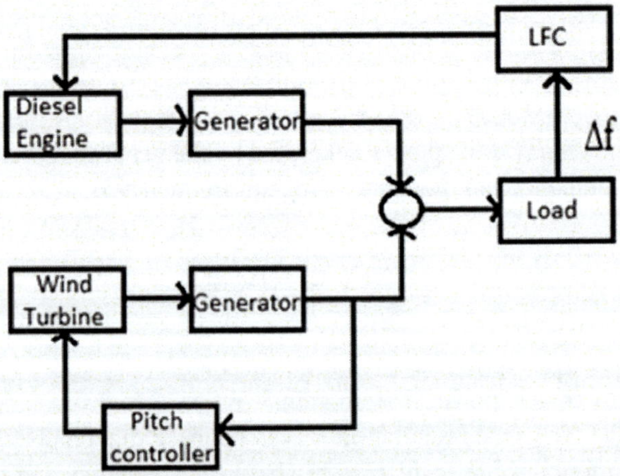

Fig. 3.3 Hybrid power system consisting of wind and diesel generating units

WTG and DEG units makes the controlling of real power generation of this hybrid power system challenging. Load frequency of this system can be controlled using proportional integral (PI) controller.

Ignoring the system losses, if P_s denotes the power generated due to wind turbine generator (P_{WTG}) and the diesel engine generator (P_{DG}), then

$$P_s = P_{WTG} + P_{DG} \tag{3.4}$$

A pitch controller is implemented to maintain the power output from the wind turbine generator during high wind velocity. This pitch controller maintains the speed of the wind turbine generator near its rated speed. The diesel engine generator system consists of a diesel engine coupled with a synchronous generator. This diesel engine generator makes the system more reliable as its output is adjusted according to the change in the load and also according to the change in the available wind power. Determining the parameters of the PI controller is also tedious. Various optimization techniques can be used for that. Two of such techniques are D-partition method and Ziegler–Nichols method. A comparative analysis shows that D-partition method reduces the frequency deviation and makes the system to settle fast [7].

3.3 Load Frequency Control of Wind-Micro Hydro-Diesel Hybrid Power Plant

A hybrid power plant shown in Fig. 3.4 consists of three generating units—wind power, micro hydro, and a diesel engine powered unit. Depending on the locations, wind and micro hydro power generations are combined with the conventional diesel generating unit. But due to environmental conditions, wind power is intermittent, which makes wind power generation variable and unpredictable and therefore needs diesel generating unit in integration with wind power generation. For expanding the hybrid power plant, a micro hydro generating unit is integrated with the existing generating units. In this hybrid power system, an alternator is connected to diesel engine, and induction generator is connected on wind turbine and hydro turbine. Reliability of the resulting power system is increased. A controller is needed to stabilize the system frequency. The controller connected to the diesel generating and the hydro turbine unit can minimize the variations in the system frequency which could be due to any change in the load or due to any change in the wind power. An adaptive neuro fuzzy interface system (ANFIS)-based neuro fuzzy controller was designed individually for both the governors; one for the diesel engine generator improves the governor performance and another for a blade pitch control in a wind turbine as a supplement controller for pitch control [8]. The system frequency variation is served as a feedback input on diesel generating unit, which further balances the differences between the generated power and the load demand by adjusting the speed changer position.

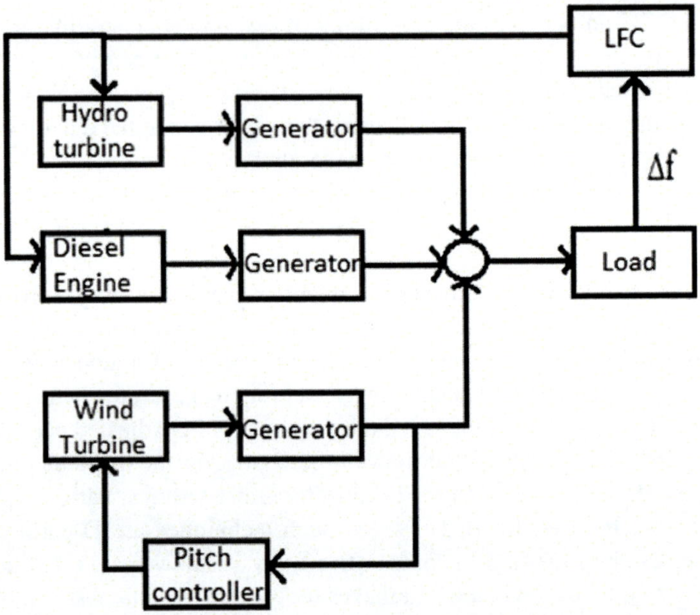

Fig. 3.4 Hybrid power system consisting of hydro, wind, and diesel generating units

3.4 Load Frequency Control of Hybrid Power System Having Energy Storage System

With the penetration of renewable energy sources into the power system, and due to its intermittent nature, energy storage systems (ESS) gains importance. There are many types of ESS like superconducting magnetic energy storage (SMES), electric batteries, fuel cells, redox flow battery, etc., which are included presently in power systems. Surplus power generated by renewable energy source (RES) can be stored in ESS and during deficit that power can be fed to the grid for the system shown in Fig. 3.5. With the increasing share of RES in the generating part, the frequency variations also increases, demanding upgrading of the load frequency controlling system. A compensation technique using ESS is the maximum power point tracking (MPPT) control of photovoltaic (PV) generation using hill climbing method, and maximum power point control tracking of wind turbine generation can be done by controlling the blade pitch. So as to maximize the output power, the output voltage of the PV system should also be controlled. PI controller is used for LFC. On adjusting the control weights of MPPT, the system frequency can be controlled [9]. SMES is having an advantage of providing immediate high power, but the power availability is for a shorter time duration. The power system is having a high percentage of RES generation and the load frequency controlling system along with SMES unit stabilizes the system frequency. The SMES is also acting as a feedback controller. In the interior of the SMES, the current in its coil needs to be restored to the initial value in a very short span after any disturbance so as to respond immediately to

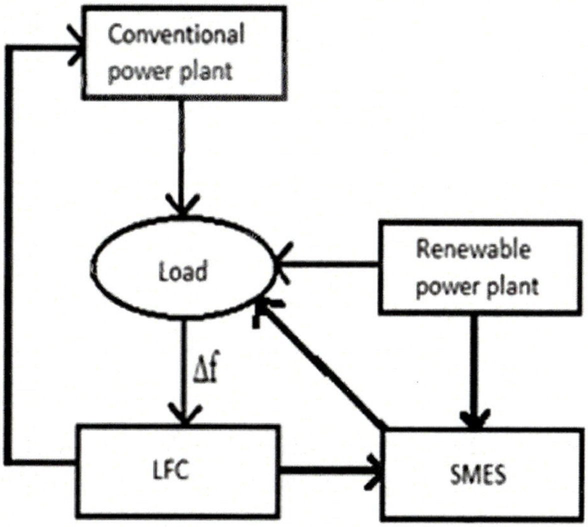

Fig. 3.5 Hybrid power system consisting of conventional, renewable power plants, and SMES

next disturbance. For this, the current change is fed as a negative feedback in the SMES control loop. Particle swarm optimization (PSO) was used for optimizing the parameters of PI controller for controlling the system frequency [10].

3.5 Load Frequency Control of Hybrid Power Plant Using V2G

On connecting electric vehicle to a power system or a grid, its battery can be used to store power. So the battery can draw power and that power can be used for driving the electric vehicle; also the power stored in the battery can be pumped back to the power system when needed. Statistical data shows that private cars are parked for most of the time, and in this duration, they can be connected to the power system. With increasing numbers of electric vehicles, the storage available for the generated power is also increasing. This technology of using electric vehicles for storing electric energy is known as V2G (vehicle to grid) technology. These electric vehicles store the surplus power generated and also feed power to the grid when needed and thereby make the generated power equal to the load and so eliminating the frequency deviations. Hybrid power system shown in Fig. 3.6 consists of diesel engine powered generator (DEG), wind turbine generator (WTG), and electric vehicle (EV) as energy storage system. Multivariable generalized predictive control (MGPC)-based load frequency controlling is used, and the control signal is generated in accordance with any change in the output of WTG or any changes in the load. The output of the controller is fed to the diesel engine generating and the electric vehicle storage which will balance the variations in the frequency. The system can be operated both in the grid connected

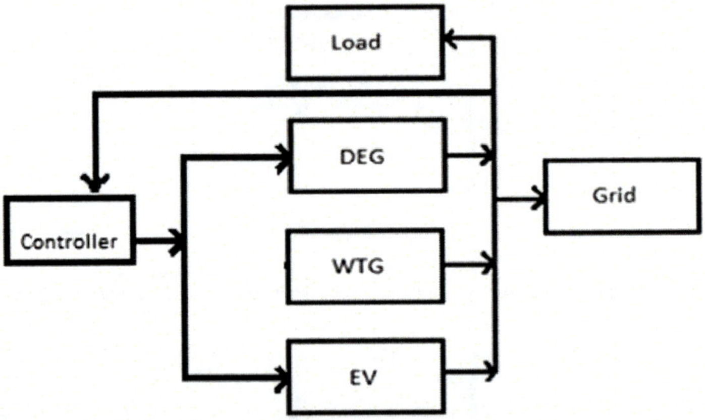

Fig. 3.6 Hybrid power system consisting of diesel, wind generating station and incorporating electric vehicle for energy storage

mode as well as islanded mode depending on the power generated and the load demand.

In the power system, the load is supplied by all three sources, and the output of the diesel engine generator and electric vehicle is controlled with respect to the change in the system frequency. Disturbance in the system is due to both the load changes as well as changes in wind power generation.

The load frequency controller used consists of two control layers. Layer 1 is controlling the individual output power of the diesel engine generator, and layer 2 is maintaining the system frequency stable. Whenever there is a change in frequency or active power, then layer 1 sends a control signal to layer 2 which further controls the output of DEG and EV so as to balance the frequency changes. Measuring units are installed with each generating units and load which measure the current state of the system. Then, forecasting is performed so as to determine the upcoming frequency changes. Then, the controller determines the control variables and sends the signal to DEG and EV [11].

A three area system having small hydro, thermal, and gas turbine units as conventional sources and solar as non-conventional source of power generation. Electric vehicles are used as ESS. On using a cascade combination of two controllers—Integral derivative with filter and another controller as a fractional order proportional derivative, the system dynamics were improved even after nonlinearities like generation rate constraint and time delay are considered [12].

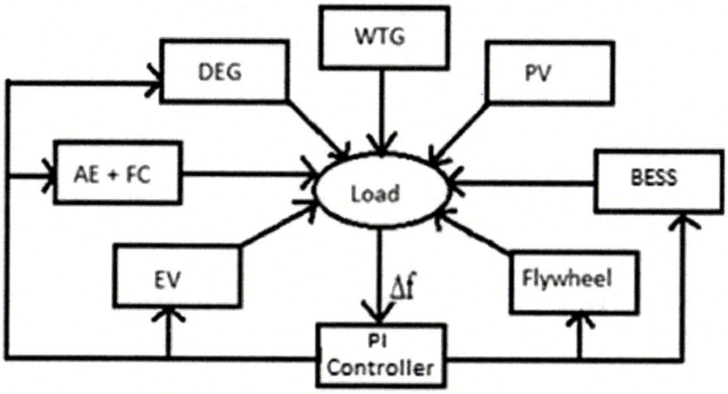

Fig. 3.7 Power system consisting of photovoltaic, wind, diesel, flywheel, fuel cell, battery, and electric vehicle

3.6 Load Frequency Control of Hybrid Power System Having Four Energy Storage Systems

A hybrid power system shown in Fig. 3.7 consists of wind turbine generator (WTG), photovoltaic generation (PV), fuel cell (FC), battery energy storage (BESS), diesel engine generator (DEG), flywheel (FW), and electric vehicle (EV). Aqua electrolyzer (AE) dissociates water into hydrogen and oxygen. Hydrogen evolved as due to electrolysis is collected in a hydrogen tank which is further used as fuel for FC. The error signal which is due to the disturbances in both the load and the renewable source generation is having both high as well as low frequency component. Battery energy storage and flywheel are fast discharging storages, so they are used to minimize fast varying components of the system frequency variations. On the other hand, fuel cell and diesel engine generator are slow discharging components and are used to control low varying components of the system frequency variations. So as to stabilize the system frequency, two proportional integral (PI) controllers are used in this power system. For optimizing the parameters of the proportional integral (PI) controllers, epsilon multi-objective genetic algorithm (ε-MOGA) is used [13].

3.7 Load Frequency Control Involving FACTS and ESS

Power semiconductor switches have revolutionized the modern power sector. They provide high efficiency, reliability, and easy controlling. These switches find applications in transmission of power as well. These systems are known as flexible AC transmission system (FACTS). FACTS devices enhance power transmission capacity of the system and manages better power flow between interconnected power systems. FACTS devices can be series connected like static synchronous series compensator (SSSC), shunt connected like static synchronous compensator (STATCOM), and

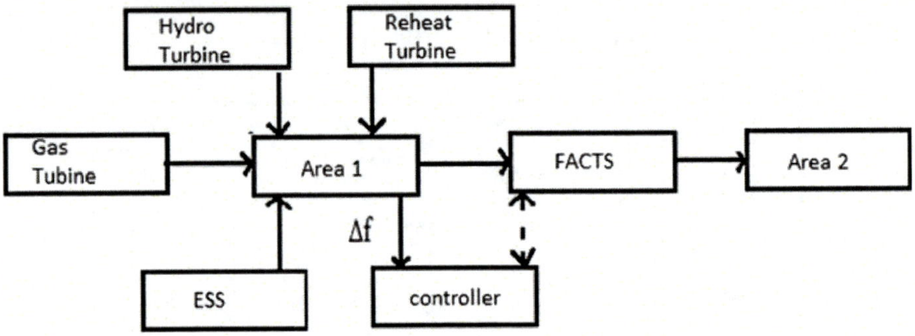

Fig. 3.8 Two-area power system incorporating FACTS

combination of series and shunt like unified power flow controller (UPFC). FACTS devices can operate in coordination in interconnected power system for controlling the deviations in frequency and improve the efficiency of power transfer through tie line thereby improving the load frequency control in case of small changes in load. Figure 3.8 shows two control areas; control area 1 is connected through a tie line to another area 2. Area 1 is having three generating units, i.e., reheat thermal turbines, hydro turbine, and gas turbine power units. Area 1 is also having energy storage system (ESS) which feeds the integral (I) controller so as to compensate for the time lag and hence maintain the system frequency within acceptable limits. Redox flow battery (RFB) is used in energy storage system. The error signal serves as the input for the RFB. SSSC is the FACTS controller used here for inducing a voltage in quadrature with line current so as to control the reactive power and enhance power transmission through tie line. SSSC when kept in insulation in series with the tie line will help in balancing frequency. SSSC can insert both inductive and capacitive reactance, and so can regulate the flow of active power through tie line. When SSSC is inserted in the tie line, it reduces the fluctuations. Higher and lower values of compensation required from SSSC can be obtained by controlling the triggering of power semiconductor switches in the SSSC. RFB eliminates the lag time hunting in the system caused by changes in load. The parameters of the integral controller are optimized using genetic algorithm (GA) [14]. The same control action can be performed using SMES as the energy storage system. Fuzzy PID controller is used for the frequency control. The parameters of the PID controller are optimized using firefly algorithm. Such control mechanism need not reset even for wide variations in the load and the system parameters [15].

3.8 Conclusion and Future Scope

Hybrid power plants offer several advantages of renewable resources and their percentage in the power generation sector needs to be increased. However, the intermittent nature of these resources cause frequency deviations. Although various types of control systems are available to deal with the frequency stability, but in future, renewable sources will penetrate more in the power sector, and so we need to develop more sophisticated load frequency control systems to tackle the issues.

References

1. D.H. Tungadio, Y. Sun, Load frequency controllers considering renewable energy integration in power system. Elsevier Energy Rep. **5**, 436–453 (2019)
2. T. Shimizu, M. Hirakata, T. Kamezawa, H. Watanabe, Generation control circuit for photovoltaic modules. IEEE Trans. Power Electron. **6**(3) (2001)
3. Z. Xilin, J. He, B. Fu, L. He, G. Xu, A system compensation based model predictive AGC method for multiarea interconnected power systems with high penetration of PV system and random time delay between different areas. Hindawi Math. Prob. Eng. **2018**(Article ID 9347878)
4. A. Jacob, L.Y. Pao, P. Flemming, E. Ela, Controlling wind turbines for secondary frequency regulation: an analysis of AGC capabilities under new performance based compensation policy, in *13th International Workshop on Large-Scale Integration of Wind Power Into Power Systems as Well as on Transmission Networks for Offshore Wind Power Plants*, Berlin, Germany (2014)
5. R.K. Sahu, T.S. Gorripotu, S. Panda, Automatic generation control of multi-area power systems with diverse energy sources using teaching learning based optimization algorithm. Eng. Sci. Technol. Int. J. **19**, 113–134 (2016)
6. Y.V. Hote, S. Jain, PID controller design for load frequency control: past, present and future challenges. IFAC papersonline **51**(4), 604–609 (2018)
7. A.J. Veronica, N.S. Kumar, F.G. Longatt, Design of load frequency control for a microgrid using D-partition method. Int. J. Emerg. Electric Power Syst. (2020)
8. S.H. Wahhab, A.K. Bhardwaj, S. Prakash, Load frequency control of hybrid system using industrial controller and implement fuzzy controller practically using PLC. Int. J. Electr. Electron. Eng. Res. (IJEEER) **3**(4), 187–204 (2013). ISSN 2250-155X
9. L. Lei, H. Matayoshi, M. E. Lotfy, M. Datta, T. Senjyu, *Load Frequency Control Using Demand Response and Storage Battery by Considering Renewable Energy Sources.* www.mdpi.com/journal/energies (2018)
10. M. Gaber, G. Shabib, A.A. Elbaset, Y. Mitani, Optimized coordinated control of LFC and SMES to enhance frequency stability of a real multi-source power system considering high renewable energy penetration. Prot. Control Modern Power Syst. **3**, 39 (2018)
11. Y. Jun, Z. Zeng, Y. Tang, J. Yan, H. He, Y. Wu, Load frequency control in isolated micro-grids with electrical vehicles based on multivariable generalized predictive theory. MDPI J. Energies 2145–2164 (2015)
12. S. Arindita, L.C. Saikia, *Renewable Energy Source-Based Multiarea AGC System with Integration of EV Utilizing Cascade Controller Considering Time Delay* (2019)
13. L.M. Elsayed, T. Senjyu, M.A.-F. Farahat, A.F. Abdel-Gawad, A. Yona, A frequency control approach for hybrid power system using multi-objective optimization. MDPI J. Energies **10**, 80 (2018)
14. S. Ravi, R. Bhushan, K. Chatterjee, Small signal stability analysis for two area interconnected power system with load frequency controller in coordination with facts and energy storage device. Ain Shams Eng. J. 603–612 (2016)
15. P.P. Chandra, R.K. Sahu, S. Panda, Firefly algorithm optimized fuzzy PID controller for AGC of multi-area multi-source power systems with UPFC and SMES. Eng. Sci. Technol. Int. J. **19**, 338–354 (2016)

Chapter 4
Time Duration Prediction of Electrical Power Outages

Rishabh Doshi, Rishabh Dev Saini, and Shivam Kansal

Abstract Electrical power outages can have negative impacts and can lead to fatal outputs that can lead to blackouts. Increased understanding and prediction of these disturbances can help to avoid the occurrence of significant disturbances and decrease the after-effects too. Previous works have focused to predict the type of electricity disturbance and their occurrences. This paper aims to develop a system that predicts the duration of power outages using machine learning and deep learning techniques. The proposed system uses preprocessing and feature selection with classification using KNN, decision tree, random forest classifiers, SVM and neural network on an open-source dataset containing electrical disturbances. The results indicate that neural networks can better predict the electrical power outage duration ranges than classic machine learning techniques.

Keywords Electrical power disturbances · Machine learning · NERC

4.1 Introduction

Electrical power systems are a crucial aspect of modern civilization. The disruptions in such systems can prove to be detrimental for businesses, governments, economies and individuals. It is a known fact that most systems are dependent on electrical power and need an uninterrupted supply of electricity. Recent developments in technology have resulted in the inception of systems that provide electricity from non-conventional sources such as solar radiation energy, wind power energy, thermal power and even nuclear power. These sources are yet to be realized to their full

R. Doshi · R. D. Saini (✉) · S. Kansal
Delhi Technological University, Delhi, India
e-mail: rishabhdevsaini_2k17co265@dtu.ac.in

R. Doshi
e-mail: rishabhdoshi_2k17co266@dtu.ac.in

S. Kansal
e-mail: shivamkansal_2k17co325@dtu.ac.in

A. Khosla and M. Aggarwal (eds.), *Smart Structures in Energy Infrastructure*,
Studies in Infrastructure and Control,
https://doi.org/10.1007/978-981-16-4744-4_4

potential and need further research and development to serve the needs of an entire society. Until then, most of us will still be dependent on conventional electrical power sources.

The disruptions in electrical power systems are of significant concern and need to be addressed. Major power disturbances are also called blackouts. Though blackouts are not very frequent, their study is still important as they have devastating effects. The main factors that cause electricity disruptions are the failure of different components of electrical power systems. These failures are initiated by a disturbance and sometimes by more than one disturbance. Exogenous non-controllable disturbances that can cause blackouts are: (1) nature disasters such as tornados, hurricanes, earthquakes, strong winds, fires and lightning; (2) human errors which are caused human mis-operation, incorrect settings of the protection devices, cyber-attacks or physical attacks; (3) unexpected/unusual component failures caused by changing operation status, failures caused by local natural vegetation or animals; (4) system failures such as a transmission line triggered by a distance relay due to overcurrent or under-voltage, voltage collapse, abnormal speed in generators, generators tripped due to under-frequency or under-voltage, instability of small-signal [1]. On account of the degree of effects of these blackouts, the study of electricity disturbances has attracted considerable attention. The Disturbance Analysis Working Group (DAWG) of the North American Electric Reliability Corporation (NERC) maintained data of electrical disturbances in North America that have occurred in the last 20 years. Developing a prediction model that can predict the duration of time for which a disturbance can persist can be of significant importance.

This will not only help the electricity company to tell the customers estimated time of blackout but can also help in repair of the fault. More relevant researches are discussed in the next section.

Previous works involving machine learning and deep learning techniques have focused on predicting the electricity disturbances according to their types [2] and exploring models to predict the weather-based electricity disturbances [3]. Cornforth [4, 5], Cornforth and Nesbitt [6] clustered the different types of electrical disturbances. Bompard et al. [7] designed a system to analyse the failures and take action toward restoration of the supply real-time. These studies have focused on the occurrence of the electricity disturbance. Our proposed work consists of a system to predict the time duration of the electricity fault, which may additionally help these systems to predict the time duration of restoration along with the type of electricity supply.

This study predicts the duration range of the electricity disturbances via various machine learning and deep learning techniques. We are particularly interested in applying MLTs to predict the range of time duration for which the disturbance persists. Several well-known developed MLTs such as K-nearest neighbours (KNN), decision trees (DT), random forest (RF), support vector machine (SVM), artificial neural networks (ANN) have been used to predict the desired result. Out of the different classifiers, the one that achieves the highest accuracy is the recommended one.

The rest of the paper is as follows: Sect. 4.2 consists of previous related works, followed by Sect. 4.3 with proposed work. Results have been discussed in Sect. 4.4 followed by Sect. 4.5 comprising conclusion and future work.

4.2 Previous Work

There are a number of studies which discuss about electrical outages and obtain different results with different methods. Omran and El Houby [2] classified the type of electricity disturbance and used four machine learning methods and ANN achieving a max accuracy of 86.11%. They used the NERC DAWG dataset. The features were manually chosen. Brester et al. [3] aimed to predict the weather-based fault prediction, with electricity fault data from Finland due to bad weather. They achieved a low mean square error for their models. MLP along with RNN, LSTM models were used. Bompard et al. [7] designed a framework to analyse the failures and take actions to maximize the restoration of unreserved loads. They used the NERC DAWG dataset. This was an attempt to simulate the behaviour of the electricity system after an unfavourable incident. Cornforth [4] was able to notice and cluster the disturbances identifying their seasonal nature with data being used for 26 years. This was done to explore the effects of machine learning techniques on the dataset. They were more focused on how the electrical disturbances can be classified into different clusters. Cornforth [5] used the clustering on large-scale disruptions and analysed the results. Owerko et al. [8] predicted the electrocal power blackouts based on different weather measure readings and using the ANNs to classify and improve the baseline prediction error. The error of 1.04% is an improvement with respect to the baseline of 3.77% error (obtained by just estimating no power outage irrespective of the weather measurements). Arif and Wang [9] analysed 6 years of distribution network outage data and predicted that the statistical properties of different outages showed the existence of a proper correlation between the repair/restoration time and the number of customers interrupted. The mean error for the restoration time predictive model was 2 h. Sun et al. [10] proposed a twitter-based prediction system using supervised latent Dirichlet allocation (sLDA) and achieved an accuracy of 81.6%. Most of the studies try to predict the electricity disturbance occurrence or the type of electricity disturbance, we propose a system to predict the duration of the electricity disturbance which can be a significant addition to this result, and this will help the consumers better understand and plan their work according to the electricity disturbance and allow electricity supply company to better manage and repair electricity disturbances.

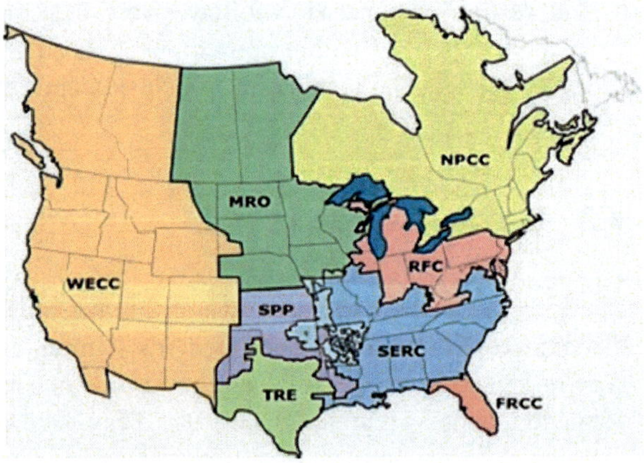

Fig. 4.1 NERC regions in North America[1]

[1] https://www.eia.gov/electricity/data/eia411/.

4.3 Dataset Details

The dataset is sourced from the EIA (US Energy Information Administration) website. This dataset contains data for the last 20 years and has been compiled by the North American Electric Reliability Corporation (NERC) Disturbance Analysis Working Group (DAWG) who compiles this data once every year. Our data contains information for all the disturbances from 2001 to 2016. The dataset is public and can be used by anyone. DAWG[1] has published some of the selected electrical disturbances reviews in North America. This has been the most widely used dataset which has been used in several studies like [2, 4, 5, 7]. The EIA dataset contains 56 features for the electrical disturbances.

We have extracted the important features that are being used for our analysis such as date and time, no of customers affected and NERC region (Fig. 4.1).

4.4 Proposed Work

This study presents methodology to predict the time duration range of electricity disturbances. Four machine learning models namely KNN, decision tree, SVM, random forest classifier along with neural networks have been discussed. The model and the training details are mentioned further. The inputs to the models are five features namely month, day, NERC region, cause category and climate category, while the output features the duration range. The proposed work consists of two phases, which are data preprocessing and classification into different time ranges based on various machine learning and deep learning models.

In the preprocessing phase, two steps are performed, which are:

1. Cleaning the raw data
2. Data formatting.

The classification phase is done using different MLTs to build different classifiers that predict the range of duration for an electricity outage.

4.4.1 Preprocessing Phase

The raw North American Electric Reliability Corporation (NERC) Disturbance Analysis Working Group (DAWG) dataset available contained 1534 data points with 52 columns. We choose eight relevant columns out of the available columns since the other columns lacked many values and had null values. The selected columns were month, day, outage start day, outage end date, NERC region, outage restoration date, cause category and climate category as were done by Cornforth [4, 5], Omran and El Houby [2]. The fields which were null in the dataset were dropped, and around 695 rows were remaining. The duration for each electricity disturbance was calculated in minutes from outage start time and outage end time.

Further, we divided the duration into partitions since all of the other data is classified as discrete and the duration of the electricity outage was continuous data, and thus, in this study, we created several buckets in the dataset with each bucket containing an equal number of values and then applied machine learning and deep learning methods in all of them.

4.4.2 Conversion of Dataset to Buckets

We have performed this study in three variants, and each time, the dataset was divided on the basis of the duration ranges. We have tried the study after dividing the duration of the electricity fault into 2–4 ranges. At every step, the no of data points were equal in each of the dataset partitions. So for two partitions, the duration range in the dataset was converted to discrete value from the continuous value. After dividing the dataset into buckets and converting the duration range from continuous to discrete, we predict the duration of the electricity ranges (Table 4.1).

Table 4.1 Representation of bucket sizes and partitions made for each bucket size

Bucket size	Partition 1 (in min)	Partition 2 (in min)	Partition 3 (in min)	Partition 4 (in min)
2	1–510	511–10,000	–	–
3	1–180	181–1560	1561–10,000	–
4	1–102	103–710	702–2880	2881–10,000

4.5 Classification Phase

The input features are taken as month, day, NERC region, cause category and climate category which will predict the range of duration of the outage. With over six features, we attempt to classify the electricity outage range with five different machine learning and deep learning techniques, viz. KNN, decision tree, random forest, SVM and artificial neural networks. The results and their probable causes are explained in the next section. Each model is validated using tenfold cross-validation. The model and the training details are mentioned in the next subsection. Figure 4.2 describes the proposed approach used for this study.

Fig. 4.2 Proposed approach for this study

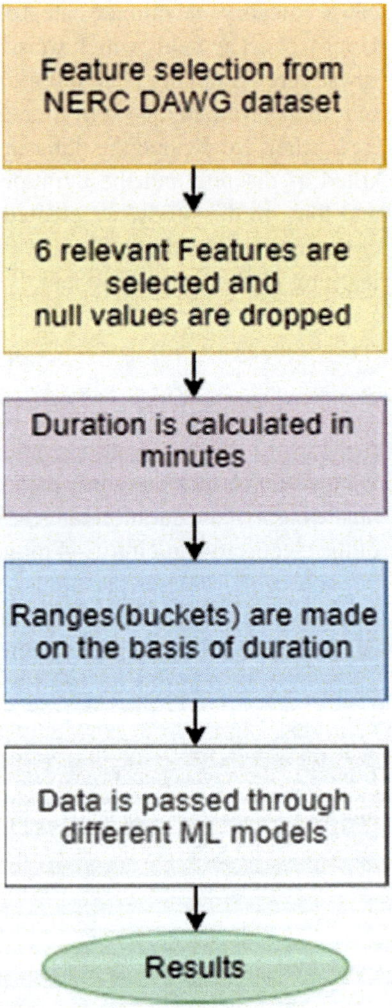

Table 4.2 Results for 2-bucket study

Method	Precision	Recall	$F1$-score
KNN	0.70	0.69	0.70
Decision tree	0.71	0.69	0.70
Random forest	0.74	0.75	0.75
SVM	0.73	0.72	0.73
Neural network	0.85	0.83	0.84

Table 4.3 Results for 3-bucket study

Method	Precision	Recall	$F1$-score
KNN	0.68	0.67	0.67
Decision tree	0.68	0.69	0.67
Random forest	0.72	0.71	0.71
SVM	0.70	0.69	0.69
Neural network	0.81	0.80	0.80

4.6 Model Details

Five machine learning techniques are used with precision, recall and F1-score as the evaluation metrics. All the techniques are validated with tenfold cross-validation. The model training details are as follows:

KNN—The best value of k came out to be 5 from grid search.
Decision tree classifier [11]—The max depth came out to be 3 from the grid search cv
Support vector machines—rbf kernel [12] was chosen. The degree was chosen as 3 and gamma was 0.01
Random forest classifier [13]—The n_estimators were kept as 50, and the max_depth was 3.
Artificial neural network—Simple dense layers were used, and the architecture was as follows
Dense(128) $\rightarrow$ Dense(64) $\rightarrow$ Dense(16) $\rightarrow$ Dense(4) $\rightarrow$ Dense(2, 3, 4).

The model was trained using the Adam optimizer [14] for 20 epochs using binary cross entropy as the loss function. L2 regularization [15] is applied in each layer and has $\lambda = 10-5$. The last layer of the neural model is changed according to the number of buckets in the dataset.

Table 4.4 Results for 4-bucket study

Method	Precision	Recall	$F1$-score
KNN	0.65	0.64	0.64
Decision tree	0.63	0.61	0.62
Random forest	0.70	0.71	0.70
SVM	0.68	0.67	0.67
Neural network	0.74	0.75	0.75

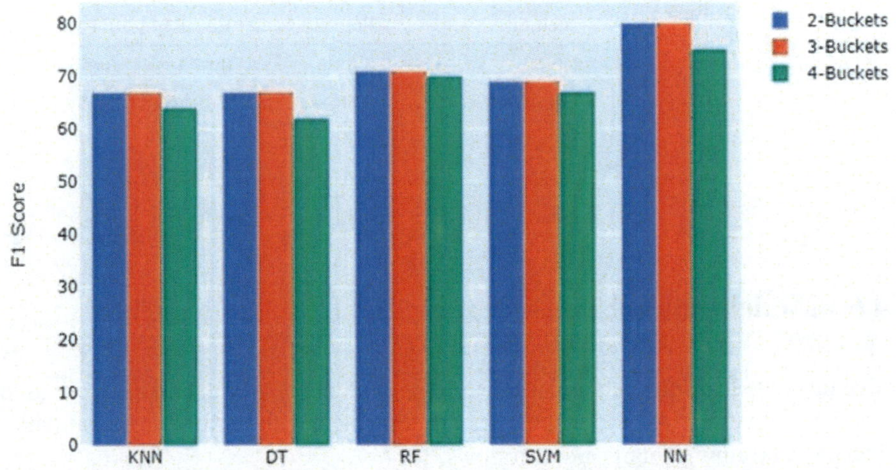

Fig. 4.3 Comparisons of different results

4.7 Experiments and Results

The results obtained with various methods are given in Tables 4.2, 4.3 and 4.4. $F1$-score, precision and recall are used as the evaluation metrics. The best results are obtained with a neural network and random forest classifier. With increasing the partitions in the dataset, the performance of the model decreases, and the neural network obtains a benchmark $F1$-score of 0.85 with two partitions and can be used for future works. Figure 4.3 compares all the results obtained.

4.8 Discussion

This paper aims to explore the effects of various machine learning methods for predicting the duration of various electrical outages. In general, neural networks perform better than all other machine learning techniques. Also, it is observed that as we increase the number of partitions of the duration of the electricity outage, the performance of the model decreases. The possible cause for this is that since the

Table 4.5 Comparison of previous works with the proposed work

Name of the paper	Year	Work done	Results obtained
Omran et al. [2]	2019	Predicted the type of electricity disturbances (5 categories)	Best accuracy of 86.11%
Brester et al. [3]	2020	Predicted the occurrence of electricity faults	Information of agreement (IA) averaged 0.68
Owerko et al. [8]	2018	Predicted the weather-based electrical disturbances	Best $F1$ score of 0.86 was obtained
Sun et al. [10]	2017	Twitter-based power outage prediction	Best accuracy of 81.6% is obtained
Our work		Predicting the duration time of the power outages	Best $F1$ score of 0.84 is observed

number of categories is increased for the machine learning and deep learning models, the models' performance would be decreasing since the feature set is limited to only five features. The performance of the neural networks can be attributed to the fact that neural networks can better catch the complex and nonlinear relationships between features and hence perform better.

Previous works [2, 4, 5] have focused on predicting the type of outages, but it is important to work further by working toward predicting other important aspects of electrical outages, which can be useful. We have attempted to predict the ranges of the duration of electrical outages, which can be useful for any future work in this field (Table 4.5).

4.9 Conclusion

This work has predicted the duration ranges of the electrical outages using machine learning techniques and the ANNs. ANNs provide the best overall accuracy. We achieved a max $F1$ score of 0.84 in 2 buckets, while achieving 0.75 in 4 buckets. This work is also limited by the dataset which can be extended to obtain more accurate and precise results. This work can act as a baseline for future works, attempting to predict electrical outages with different datasets.

References

1. O.P. Veloza, F. Santamaria, Analysis of major blackouts from 2003 to 2015: classification of incidents and review of main causes. Electr. J. **29**(7), 42–49 (2016)
2. S. Omran, E.M. El Houby, Prediction of electrical power disturbances using machine learning techniques. J. Ambient Intel. Human. Comput. 1–17 (2019)
3. C. Brester, H. Niska, R. Ciszek, M. Kolehmainen, Weather-based fault prediction in electricity networks with artificial neural networks, in *2020 IEEE Congress on Evolutionary Computation*

(CEC), pp. 1–8 (IEEE, 2020)

4. D. Cornforth, Long tails from the distribution of 23 years of electrical disturbance data, in *2009 IEEE/PES Power Systems Conference and Exposition*, pp. 1–8 (IEEE, 2009)
5. D. Cornforth, Applications of data mining to time series of electrical disturbance data, in *2009 IEEE Power and Energy Society General Meeting*, pp. 1–8 (IEEE, 2009)
6. D. Cornforth, K. Nesbitt, Quality assessment of clusters of electrical disturbances: a case study, in *2013 IEEE 8th Conference on Industrial Electronics and Applications (ICIEA)*, pp. 247–254 (IEEE, 2013)
7. E. Bompard, A. Estebsari, T. Huang, G. Fulli, A framework for analyzing cascading failure in large interconnected power systems: a post-contingency evolution simulator. Int. J. Electr. Power Energy Syst. **81**, 12–21 (2016)
8. D. Owerko, F. Gama, A. Ribeiro, Predicting power outages using graph neural networks, in *2018 IEEE Global Conference on Signal and Information Processing (GlobalSIP)*, pp. 743–747 (IEEE, 2018)
9. A. Arif, Z. Wang, Distribution network outage data analysis and repair time prediction using deep learning, in *2018 IEEE International Conference on Probabilistic Methods Applied to Power Systems (PMAPS)*, pp. 1–6 (IEEE, 2018)
10. H. Sun, Z. Wang, J. Wang, Z. Huang, N. Carrington, J. Liao, Data-driven power outage detection by social sensors. IEEE Trans. Smart Grid **7**(5), 2516–2524 (2016)
11. J.R. Quinlan, Induction of decision trees. Mach. Learn. **1**(1), 81–106 (1986)
12. K. Thurnhofer-Hemsi, E. López-Rubio, M.A. Molina-Cabello, K. Najarian, Radial basis function kernel optimization for support vector machine classifiers (2020). arXiv preprint arXiv: 2007.08233
13. A. Liaw, M. Wiener, Classification and regression by random forest. R News **2**(3), 18–22 (2002)
14. D.P. Kingma, J. Ba, Adam: A Method for Stochastic Optimization (2014). arXiv preprint arXiv: 1412.6980
15. J. Schmidhuber, Deep learning in neural networks: an overview. Neural Netw. **61**, 85–117 (2015)

Chapter 5
Comparative THD Analysis of Multi-level Inverter Using SPWM Scheme

Shashi Shekhar Tripathi, Manoj Kumar Kar, and A. K. Singh

Abstract Multi-level inverter is an emerging technology in high-power and medium-voltage application. It is more superior than two-level inverters because of its operation capability at higher voltage, reduced voltage derivative, and enhanced efficiency. The overall voltage profile and efficiency of the system may be enhanced using MLI. The voltage stress is greatly reduced, and hence, the output waveform approaches to be sinusoidal with the use of MLI. In this work, CHBMLI topology is chosen for analyzing THD of different MLI. The harmonic for three-, five-, seven-, nine-, eleven-, and thirteen-level output voltage waveforms are studied and compared using MATLAB Simulink. The pulse for switches is generated using SPWM techniques. With the increased of number of levels in output waveform, THD reduces. This paper presents the improvement in THD with the increased number of levels.

Keywords Cascaded H-bridge MLI · MI · SPWM · THD

5.1 Introduction

Two new MLI topologies for staircase output voltage and reduced switch count, first one having 3 dc voltage source and 10 switches giving 15-level output and second one having 4 dc voltage source and 12 switches giving 25-level output are discussed in [1]. An advanced hybrid algorithm named APSO-NR based on SHEPWM for cascaded H-bridge MLI to eliminate unwanted lower order harmonics is discussed in [2] applicable for all levels of MLI. Modified MLI topology seven-level pack u-cell inverter for seven-level output voltage with lower harmonic distortion for photovoltaic (PV) application is discussed in [3]. Based on input dc voltage sources cascaded H-bridge MLI has two configurations symmetric (equal magnitude) and asymmetric (unequal magnitude). For same higher levels, output asymmetric configuration has less THD as per IEEE standard than symmetric configuration [4]. Highly frequency magnetic linked (HFML)-based CHB MLI has advantages of reduced device count, size, and

S. S. Tripathi (✉) · M. K. Kar · A. K. Singh
NIT Jamshedpur, Jamshedpur, Jharkhand, India

A. Khosla and M. Aggarwal (eds.), *Smart Structures in Energy Infrastructure*,
Studies in Infrastructure and Control,
https://doi.org/10.1007/978-981-16-4744-4_5

cost than conventional invertors is discussed in [5]. The cascaded H-bridge topology of MLI is more efficient than diode-clamped and capacitor-clamped MLI. As no. of output voltage levels in cascaded H-bridge MLI increases, the harmonic distortion and voltage stress d*v*/d*t* reduces [6]. A new MLI topology using phase disposition (PD) SPWM techniques for high-voltage and high-power application is discussed in [7, 8]. Latest advancement in MLI topologies, modulation, and control techniques are discussed in [9].

Symmetric and asymmetric topologies of MLI generating all levels of output voltage with reduction in switches, cost, and size are discussed in [10]. A current control algorithm using discrete time model for cascaded H-bridge MLI generating all levels of output voltage are discussed in [11]. Particle swarm optimization (PSO) method to solve harmonic elimination problems with unequal dc sources in cascaded H-bridge MLI is discussed in [12]. Series connection of cascaded H-bridge cells with photovoltaic (PV) module for tracking the maximum power point is discussed in [13]. Three-phase cascaded H-bridge MLI circuit formation and various types of MSPWM techniques are discussed in [14]. Emerging topologies, most relevant control, and modulation methods and latest applications of MLI are discussed in [15].

5.2 Cascaded H-bridge MLI

The basic structure of CHBMLI is H-bridge. It is also called cells or modules. The CHB has many cells connected in series. The dc link of each cells is isolated. With m cells no. of levels in output waveform is $(2\,m + 1)$. In some cases, for same output voltage there are many switching states possible, it is called multiplicity. Multiplicity in switching states is an inherent property. It can be used for redistributing the losses or charge/discharge capacitors.

For $m = 2$, CHBMLI has five levels of voltage output $2E$, $-\,2E$, E, $-\,E$, and 0, respectively. Its switching table and circuit diag. is depicted below in Fig. 5.1 (Table 5.1).

5.3 Problem Definition

The two-level inverter has more distortion which degrades the overall efficiency of the system. Therefore, in order to obtain better waveform, filtering circuit is required. To overcome the aforesaid problem, MLI is generally used. Nowadays, MLI gets attention due to its high-power application in industry. The different topologies of MLI have been implemented in literatures to improve the performance. In this paper, CHBMLI is proposed due to the lesser number of components requirement in comparison with other topologies. The SPWM technique has been implemented for generating pulses.

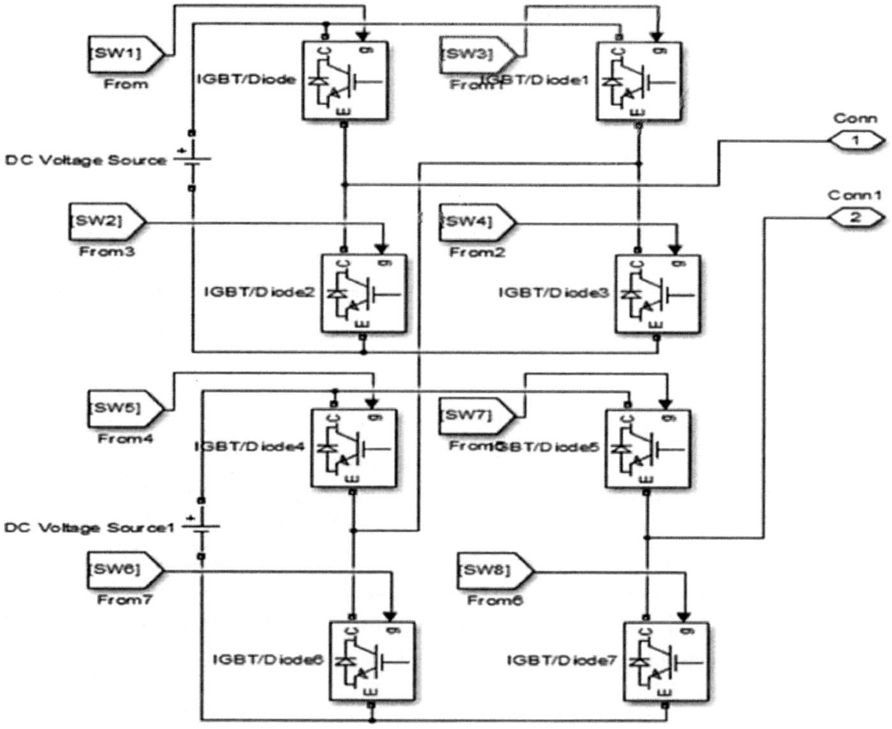

Fig. 5.1 Basic diagram of five-level CHBMLI

Table 5.1 Output vol. for five-level CHB MLI

Voltage	Sw11	Sw22	Sw33	Sw44	Sw55	Sw66	Sw77	Sw88
$2E$	1	0	0	1	1	0	0	1
E	1	0	0	1	1	0	1	0
0	0	0	0	0	0	0	0	0
$-E$	0	1	1	0	1	0	1	0
$-2E$	0	1	1	0	0	1	1	0

5.4 SPWM Scheme

PWM is the inverter's internal control tool. A constant dc input voltage is given to an inverter in order to obtain a regulated output voltage by varying the cycles (on and off) of the components of the inverter. The width of pulses in SPWM is a sinusoidal function of the pulse's angular location in a cycle. A triangular carrier wave of high frequency is compared with a reference (sinusoidal) wave of lower frequency to obtain the pulses.

Both the waves, i.e., reference and carrier waves, are compared using a comparator. When the sinusoidal wave is greater than the triangular wave, the output of the comparator is high, otherwise it is low. In a trigger pulse generator, the comparator output is processed in such a way that the output voltage wave of the inverter has a pulse width in accordance with the comparator output pulse width.

The ratio of the peak value of the reference wave to the peak value of carrier wave termed as the modulation index (MI) and MI influences the harmonic spectra of the output voltage. The magnitude of the output voltage's fundamental variable is proportional to MI. The output voltage is varied by varying the MI.

5.5 Simulation Result

In this paper, total harmonic distortion of CHB MLI topology for 3, 5, 7, 9, 11, and 13 levels output voltage waveform is studied using MATLAB simulation. IGBT is implemented as a switch. Every h-bridge module is made of four IGBT switches. Pulses are generated by SPWM techniques and provided to switches. By connecting h-bridge cells in cascaded manner, we can get desired output levels.

Each cell has input dc voltage $E = 100$v. For three-level output, total input dc vol. is 100 V and THD is 55.88%. For five-level output, total input dc vol. is 200v and THD is 26.89% depicted in Figs. 5.2 and 5.3. For seven-level output, total input dc vol. is 300v and THD is 16.43%.

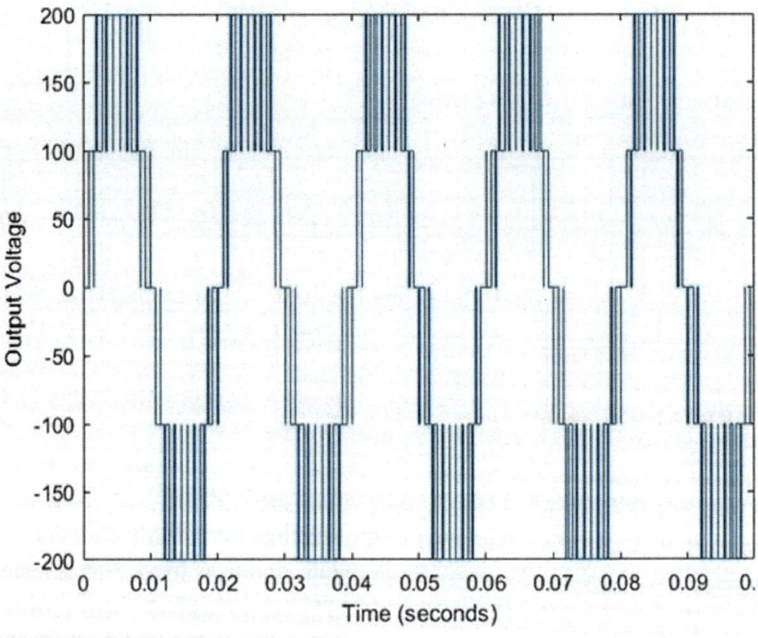

Fig. 5.2 Output voltage of five-level CHBMLI

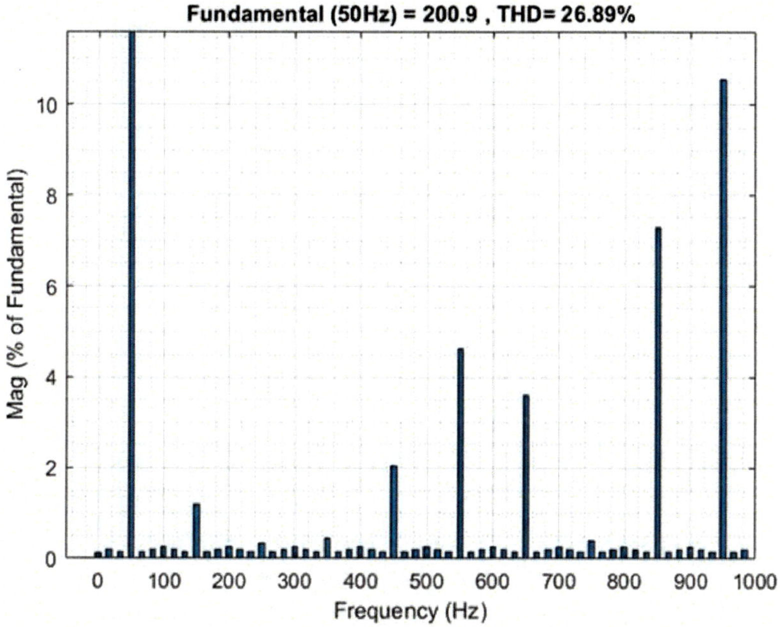

Fig. 5.3 THD percentage of five-level CHBMLI

For nine-level output, total input dc vol. is 400v and THD is 16.79%. For 11-level output, total input dc vol. is 500v and THD is 14.04%. For 13-level output, total input dc vol. is 600v and THD is 8.58% as shown in Figs. 5.4 and 5.5. The modulation index is always kept one.

5.6 Conclusion

CHB topology of multi-level inverter is preferred over neutral point clamped and flying capacitor MLI topology because it requires less power semiconductor components than other topology. The THD for 3-, 5-, 7-, 9-, 11-, and 13-level output voltages are 55.88%, 26.89%, 16.43%, 16.79%, 14.04%, and 8.58%, respectively, as given in Table 5.2. It is evident from the comparison given in Table 5.2, that as number of levels in output waveform increases, the harmonic reduces and output waveform becomes smoother and more sinusoidal.

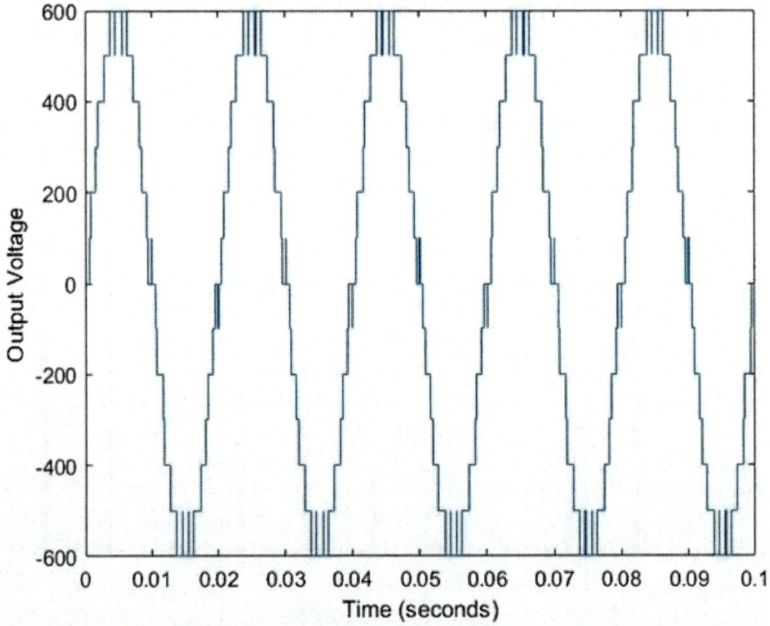

Fig. 5.4 Output voltage of 13-level CHBMLI

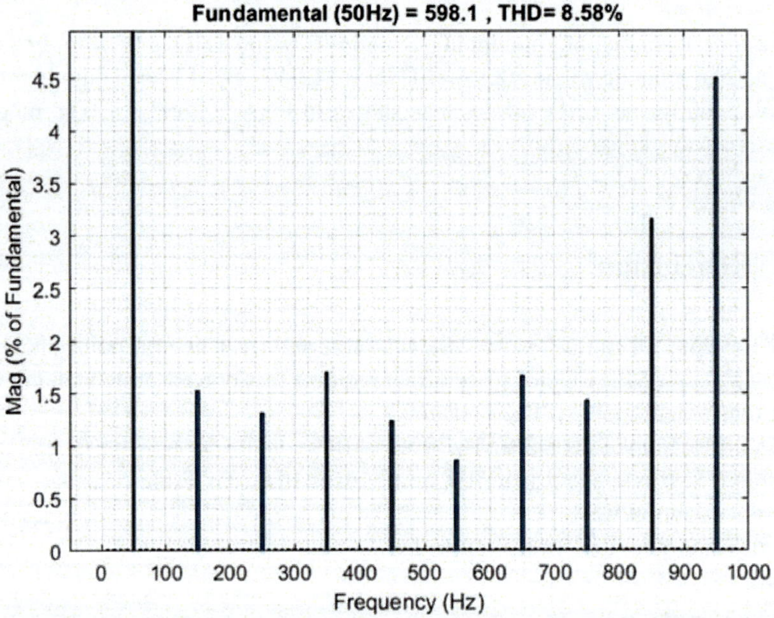

Fig. 5.5 THD percentage of 13-level CHBMLI

Table 5.2 THD percentage of different level

Serial no.	No. of levels	THD percentage (%)
1	3	55.88
2	5	26.89
3	7	16.43
4	9	16.79
5	11	14.04
6	13	8.58

References

1. M.D. Siddique, A new multilevel inverter topology with reduce switch count. IEEE Access **7**, 55584–55594 (2019)
2. M.A. Menon, S. Mekhilef, M. Mubin, Selective harmonic elimination in multilevel inverter using hybrid APSO algorithm. IET Power Electron. **11**(10), 1673–1680 (2018)
3. H. Vahedi, M. Sharifzadeh, K. Al-Haddad, Modified seven level pack u-cell inverter for photovoltaic application. IEEE Trans. Merg. Sel. Top. Power Electron. **6**(3), 1508–1516 (2018)
4. C. Dhanamjayulum, S. Meikandasivam, Implementation and comparision of symmetric and asymmetric multilevel inverters for dynamic loads. IEEE Access **6**, 738–746 (2018)
5. M. Hasan, A. Abu-siada, S.M. Islam, M. Dahidah, A new cascaded multilevel inverter topology with galvanic isolation. IEEE Trans. Ind. Appl. (2018)
6. G. Singh, V.K. Garg, THD analysis of cascaded h-bridge multilevel inverter, in *4th IEEE International Conference on Signal Proccessing, Computing, and Control* (2017)
7. E. Najafi, A.H. Yatim, Design and implementation of a new multilevel inverter topology. IEEE Trans. Ind. Electron. **59**(11), 4148–4154 (2012)
8. M.K. Kar, P. Giri, N.K. Gupta, A.K. Singh, Control of a three-phase diode clamped multilevel inverter using phase disposition modulation scheme, ed. by J. Kumar, P. Jena. *Recent Advances in Power Electronics and Drives. Lecture Notes in Electrical Engineering*, vol. 707 (Springer, Singapore, 2021). https://doi.org/10.1007/978-981-15-8586-9_6
9. S. Kouro, M. Malinowski, J. Pou, J.I. Lenon, L.G. Franquelo, A. Perez, Recent advances and industrial applications of multilevel converters. IEEE Trans. Ind. Electron. **57**(8), 2553–2580 (2010)
10. R.A. Ahmed, S. Mekhilef, H. Ping, New multilevel inverter topology with minimum no. of switches. IEEE Tencon, 1862–1867 (2010)
11. P. Cortes, A. Wilson, J. Rodriguez, H. Abu-Rub, Model predictive control of multilevel cascaded h-bridge inverters. IEEE Trans. Ind. Electron. **57**, 2691–2699 (2010)
12. H. Taghizadeh, M. Tarafdar, Harmonic elimination of cascade multilevel inverters with nonequal dc sources using particle swarm optimization. IEEE Trans. Ind. Electron. **57**(11), 3678–3684 (2010)
13. E. Villanneva, P. Correa, J. Rodriguez, M. Pacas, Control of a single-phase cascaded h-bridge multilevel inverter for grid connected photo-voltaic systems. IEEE Trans. Ind. Electron. **56**(11), 4399–4406 (2009)
14. M.K. Kar, M. Mansoori, S. Kumar, S.K. Gupta, Harmonic elimination of a T-type multilevel inverter based on multistate switching cell, ed. by J. Kumar, P. Jena. *Recent Advances in Power Electronics and Drives. Lecture Notes in Electrical Engineering*, vol. 707 (Springer, Singapore, 2021). https://doi.org/10.1007/978-981-15-8586-9_9
15. J. Rodriguez, J. Slai, F.Z. Peng, Multilevel inverters: a survey of topologies, control, and applications. IEEE Trans. Ind. Electron. **49**(4), 724–738 (2002)

Chapter 6
Power Quality Improvement of Railway Traction System Using D-STATCOM

Ruma Sinha and H. A. Vidya

Abstract Electric railway traction system is a dynamically varying load with high power rating. In addition to this, the usage of power electronic converter in this system causes power quality problems like harmonic distortion, negative sequence component of current, voltage unbalance, voltage fluctuation, etc. This affects the upstream power supply and needs to be compensated. A D-STATCOM based system is proposed in order to minimize the harmonics generated in the traction power supply system and to eliminate the negative sequence component of current.

Keywords Electric traction · D-STATCOM · Power quality · THD · NSC

6.1 Introduction

Railway electrification has become popular due to multiple reasons, for example, reduced air pollution, capability to carry heavy load, reduced carbon dioxide generation, higher efficiency, etc. Railway traction uses AC or DC power supply with different voltage levels in different countries. In India, mainly 25 kV AC is used for mainline traction, 750 V DC is used in Bangalore and Kolkata Metro, while Chennai, Mumbai, and Delhi metro use 25 kV AC through overhead catenary. For a 25 kV AC system, in the traction substation, a three-phase to one-phase or two-phase traction transformer is used. Dynamic characteristic of traction load leads to draw unbalanced current and that introduces a negative sequence component which could be as high as positive sequence component [1, 2]. As the traction load is a 1-ϕ load, the two phases generated from this three-phase system will not be balanced, and also due to the characteristic of the specially connected transformer, the supply current drawn from the source becomes unbalanced which affects the power quality (PQ). This unbalance in current produces a "negative sequence component of current" (NSC). Higher NSC injecting into the grid may have several impacts on the utility grid. Increased NSC causes overall overheating of the system which reduces the lifespan

R. Sinha (✉) · H. A. Vidya
Global Academy of Technology, Bangalore, India

A. Khosla and M. Aggarwal (eds.), *Smart Structures in Energy Infrastructure*,
Studies in Infrastructure and Control,
https://doi.org/10.1007/978-981-16-4744-4_6

of the equipment connected to it. Also, protective devices may malfunction with increased NSC.

Traction loads have high power ratings with dynamic and nonlinear load characteristics. To supply the required power with proper control to the traction motor, the drive uses power electronics converters [3]. The traction drive is a variable speed drive which may require frequent start and stop. Also, the peak load varies largely with valley load. The power quality issues generated because of the dynamic characteristics of traction system affects the upstream power supply network.

6.2 Power Quality Issues in Railway Traction System

The major power quality issues in railway traction system are: system imbalance, harmonic distortion, reactive power, and voltage unbalance.

System Imbalance—This is considered as most serious issue with reference to the "power quality" of electric railway traction. Single-phase traction load produces a "negative sequence component" (NSC) as high as "positive sequence component." The traction load is of heavy duty, so this large magnitude NSC is extremely harmful for the power network and requires mandatory attenuation.

Harmonics—DC traction system uses "12 pulse rectifiers" and generates huge amount of harmonics while AC traction system uses AC-DC-AC converters, which produces harmonics of different order that flows into the 3-ϕ power systems. Current and voltage harmonics generated creates a major power quality issue and must be compensated.

Reactive Power—Power converters connected with traction drives generally use "pulse width modulation" techniques, which generate zero "reactive power" [3]. In traction substation (TSS), active and reactive power generation or transfer is necessary for the compensation of "negative sequence components." Hence, compensation of reactive power is not taken care of separately.

Voltage Unbalance—Unbalanced current produces unbalanced voltages. As per IEC 6850 and EN 50163 in traction system, motor and other loads function properly with 24% reduced magnitude and 10% increased magnitude. So, usually voltage unbalance is not considered as a serious problem.

Power quality issues may cause malfunctioning of protective relays, instabilities, reduced lifetime, failure of equipment, decreased utilization factor, etc. Hence, it becomes an absolute necessity to mitigate the power quality issues of the traction system [4]. To mitigate these power quality issues, different schemes have been proposed by researchers. Passive filters with different structure have been proposed, but the passive filters are limited to selective harmonic elimination, fixed compensation, and increased size and may lead to resonance [5, 6].

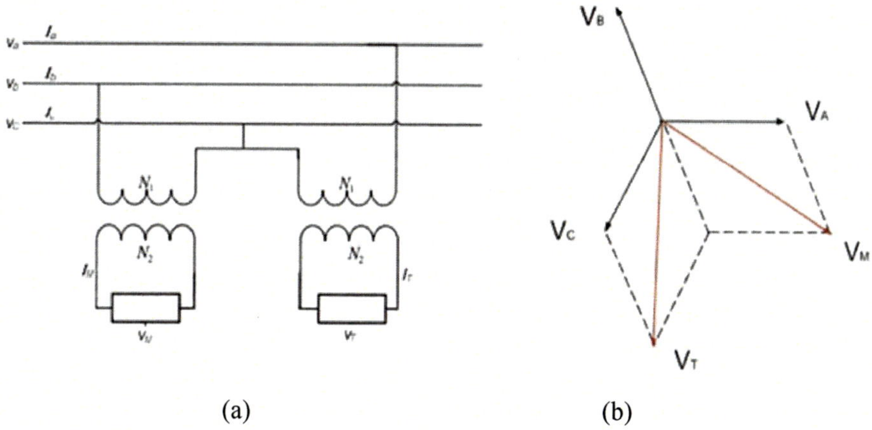

Fig. 6.1 V-V connected transformer **a.** connection diagram, **b.** phasor diagram

Specially connected transformers are proposed to supply the traction system which would reduce the total harmonic distortion [7–9]. With balanced impedance-matching transformers, no fundamental NSC is produced if load is balanced along the line. But in practice, the traction load is rarely balanced due to the number of locomotives, speed, and location. In this paper, a V-V transformer connection is used to obtain two single phases from the three-phase grid. A "Distribution Static synchronous compensator" (D-STATCOM)-based active power flow control is proposed to compensate the grid side "negative sequence current" from the traction power supply system [10–13]. Synchronous reference frame theory-based controller is implemented using MATLAB SIMULINK environment. As a V-V connected transformer has maximum power rating utilization ratio and a simple topology, V-V connected transformer is chosen for the analysis in this paper. Also the result is compared with a Scott connected transformer. The connection diagram and phasor diagram is shown in Fig. 6.1.

A block diagram for the overall system is given in Fig. 6.2 and the control strategy of D-STATCOM is depicted in Fig. 6.3. DC link voltage of the capacitor (V_d) and "Point of Common Coupling" (PCC) voltage (V_a, V_b, V_c) are taken as feedback for control.

The load current in *a-b-c* phases (I_{La}, I_{Lb}, I_{Lc}) is converted to "*d-q*-0" frame of reference using "Park's transformation" as given in Equations 6.1-6.3.

$$i_{Ld} = \frac{2}{3} i_{La} \cos \omega t - \frac{2}{3} i_{Lb} \sin \omega t + \frac{1}{3} i_{Lc} \tag{6.1}$$

$$i_{Lq} = \frac{2}{3} i_{La} \cos\left(\omega t - \frac{2\pi}{3}\right) - \frac{2}{3} i_{Lb} \sin\left(\omega t - \frac{2\pi}{3}\right) + \frac{1}{3} i_{Lc} \tag{6.2}$$

$$i_{L0} = \frac{2}{3} i_{La} \cos\left(\omega t + \frac{2\pi}{3}\right) + \frac{2}{3} i_{Lb} \sin\left(\omega t + \frac{2\pi}{3}\right) + \frac{1}{3} i_{Lc} \tag{6.3}$$

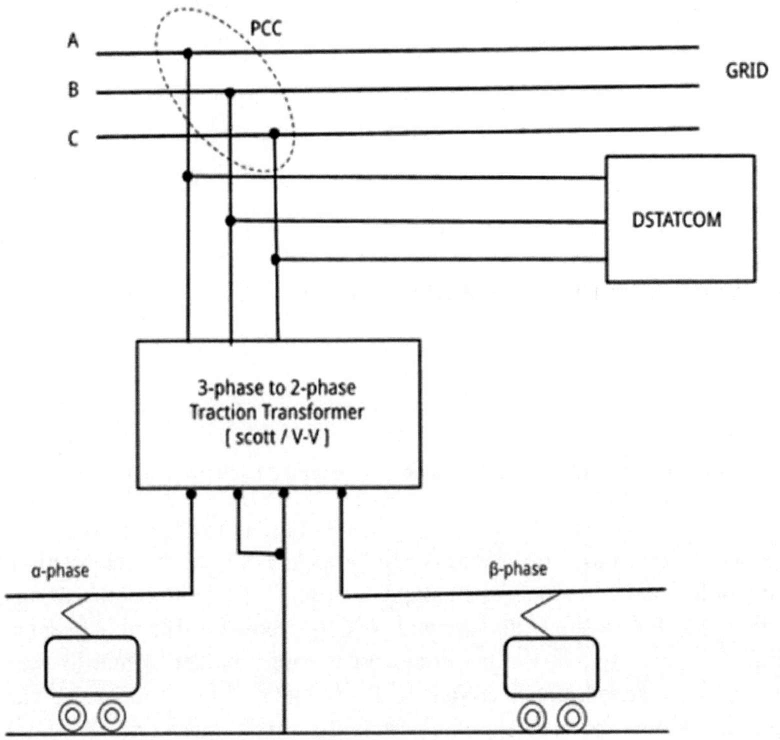

Fig. 6.2 Block diagram of the overall system

Load synchronization of the signals with PCC is achieved with a three-phase Phase Locked Loop (PLL). Currents i_{Ld} and i_{Lq} contain AC as well as DC components. Extraction of DC component is obtained by using a low pass filter (LPF).

$$i_{\mathrm{Ld}} = i_{\mathrm{dDC}} + i_{\mathrm{dAC}} \tag{6.4}$$

$$i_{\mathrm{Lq}} = i_{\mathrm{qDC}} + i_{\mathrm{qAC}} \tag{6.5}$$

$$i_{\mathrm{d\text{-}ref}} = i_{\mathrm{dDC}} + i_{\mathrm{loss}} \tag{6.6}$$

The loss component of current is obtained by using a proportional plus integral (PI) controller with the error signal of DC link voltage. The DC link voltage is obtained as

$$V_{\mathrm{d}} = \text{factor of safety} \times \frac{V_{\mathrm{LL}}}{0.707\ \mathrm{m}} \tag{6.7}$$

where V_{LL} is the line to line voltage and m is the modulation index.

The magnitude of the voltage at grid side is calculated as

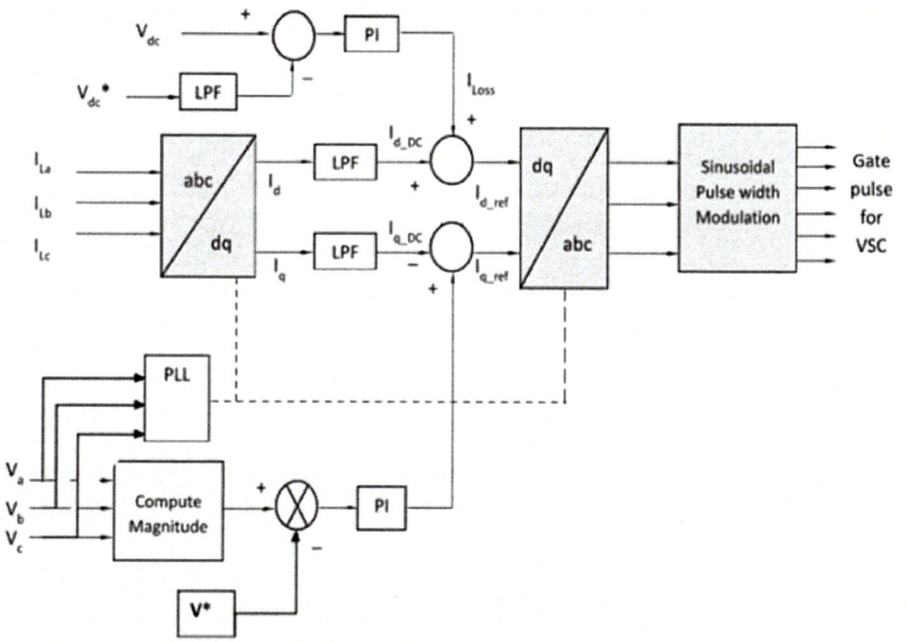

Fig. 6.3 Control strategy of D-STATCOM

$$V_{sp} = \sqrt{\frac{2}{3}(V_a^2 + V_a^2 + V_c^2)} \tag{6.8}$$

Reference current generated requires to be in phase with the supply voltage, but zero sequence component set to zero. This is obtained by reverse park's transformation with $i_{0\text{-ref}}$ equal to zero (Fig. 6.3).

$$i_{a-\text{ref}} = i_{d-\text{ref}} \cos \omega t + i_{q-\text{ref}} \sin \omega t + i_{0-\text{ref}} \tag{6.9}$$

$$i_{b-\text{ref}} = i_{d-\text{ref}} \cos\left(\omega t - \frac{2\pi}{3}\right) + i_{q-\text{ref}} \sin\left(\omega t - \frac{2\pi}{3}\right) + i_{0-\text{ref}} \tag{6.10}$$

$$i_{c-\text{ref}} = i_{d-\text{ref}} \cos\left(\omega t + \frac{2\pi}{3}\right) + i_{q-\text{ref}} \sin\left(\omega t + \frac{2\pi}{3}\right) + i_{0-\text{ref}} \tag{6.11}$$

6.3 Simulation Result

The simulation model is developed using MATLAB SIMULINK environment as depicted in Fig. 6.4. A VSC based D-STATCOM is implemented using synchronous reference frame controller theory which is installed at the grid side. The performance

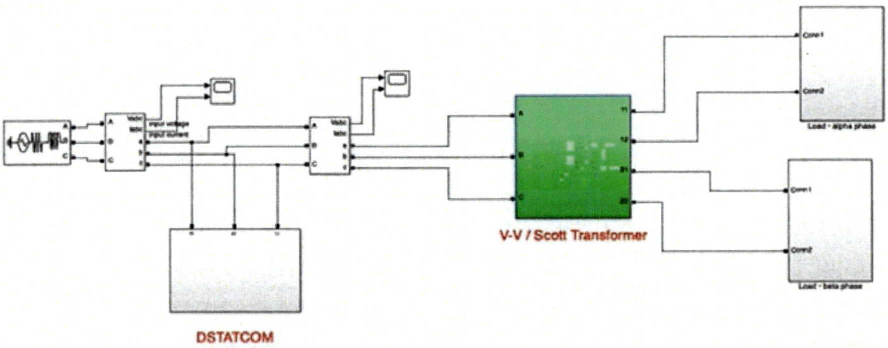

Fig. 6.4 SIMULINK model of the system with D-STATCOM controller

is analyzed for both Scott and V-V connected transformers with different load conditions. Analysis has been carried out with balanced load in both α and β phase and also with unbalanced load (Fig. 6.5).

"Current unbalance factor" is given as the ratio of negative sequence component to positive sequence component of current. "Load balance degree" is defined as the ratio of current in the light load side to the current flowing in the heavy load side.

$$\text{current unbalance factor} = \frac{\text{negative sequence component of current}}{\text{Positive sequence component of current}} \tag{6.12}$$

$$\text{Load balance degree} = \frac{|I|_{\text{light-load-side}}}{|I|_{\text{heavy-load-side}}} \tag{6.13}$$

Simulation results show that total harmonic distortion (THD) is above permissible limit in V-V connected transformer, but Scott connected traction transformer does not inject harmonics in the input side. However, the current drawn from the supply

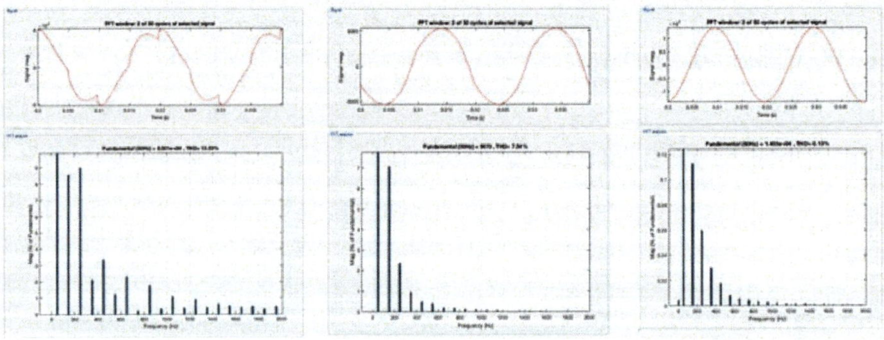

(a) voltage THD (b) Current THD (c) current THD

Fig. 6.5 Input side voltage and current THD of V-V system **a**, **b** uncompensated system, **c** compensated system is not balanced, and both the transformer connection produces negative sequence component of current. Also, with the increase in unbalance in the load side, the negative sequence component rises. This unbalance in the current drawn from the main grid can be compensated by the implementation of D-STATCOM in addition to the reduction of THD in both source current and voltage (Figs. 6.6, 6.7, 6.8 and Table 6.1).

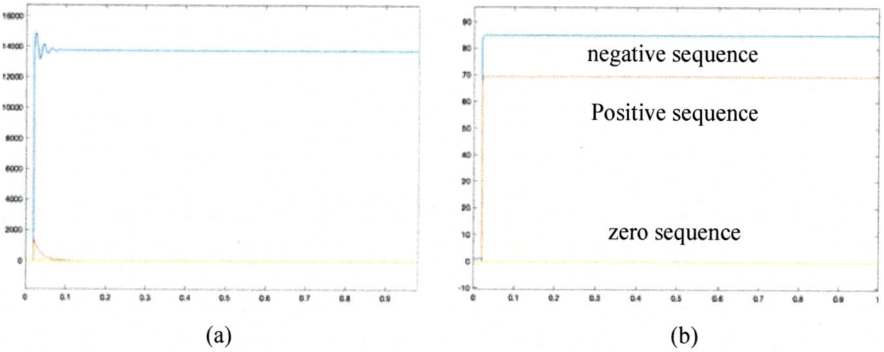

Fig. 6.6 Positive, negative, and zero sequence component of current **a** with compensation **b** without compensation

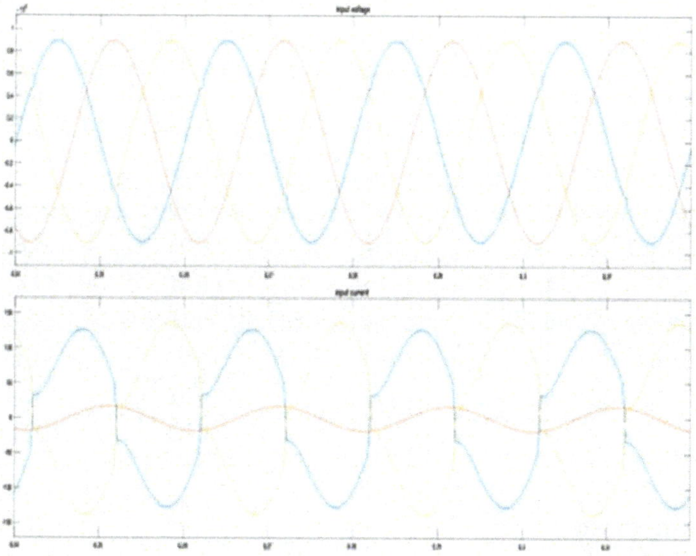

Fig. 6.7 Grid side voltage and current waveform without compensation [unbalanced load]

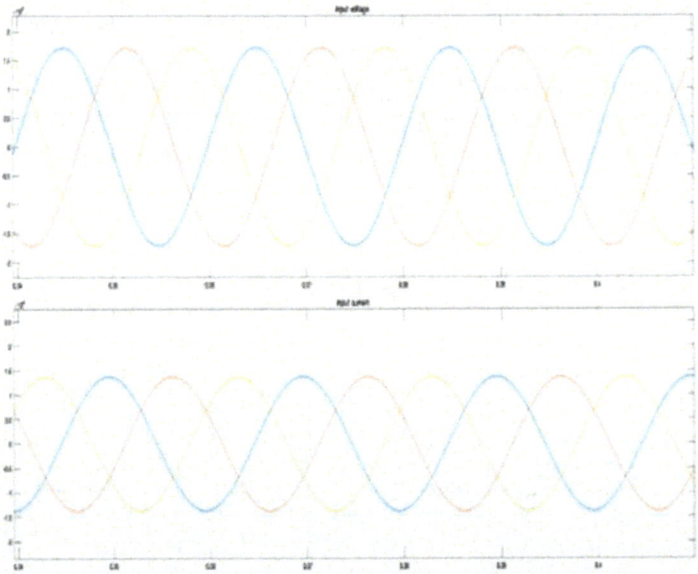

Fig. 6.8 Grid side voltage and current waveform after compensation [unbalanced load]

Table 6.1 THD and current unbalance

Transformer connection and load	Voltage THD (%)	Current THD (%)	Load balance degree	Current unbalance factor (%)
Uncompensated (Scott)	0.01	0.01	0.99	58.9
	0.01	0.01	0.19	72.77
Compensated (Scott)	0.43	0.02	0.99	0.01
	0.43	0.02	0.19	0.01
Uncompensated (V-V)	13.53	7.54	0.95	35.5
	7.42	6.25	0.51	67.4
	7.46	3.86	0.16	91.8
Compensated (V-V)	2.58	0.13	0.97	1.15
	1.36	0.11	0.53	1.26
	2.69	0.13	0.23	1.3

6.4 Conclusion

In this paper, a D-STATCOM controller is proposed to compensate the unbalance and THD in the current drawn from the main grid. The performance is analyzed for V-V and Scott connected transformer systems with balanced and unbalanced load conditions. The controller designed, without any modification, can compensate the current unbalance, NSC, and THD for the system irrespective of the load balance degree and applicable both for Scott and V-V connected transformers.

References

1. D. Zhang, D. Zhang, Z. Zhang, W. Wang, Y. Yang, Negative sequence current optimizing control based on railway static power conditioner in V/V traction power supply system. IEEE Trans. Power Electron. **31**(1) (2016)
2. G.K. Allaboyena, Mitigation of NSC and alleviate of magnetic inrush currents in Indian traction system. Int. J. Sci. Res. (IJSR) **4**(11) (2015)
3. A.B. Kebede, G.B. Worku, Power electronics converter application in traction power supply system. Am. J. Electr. Power Energy Syst. **9**(4), 67–73 (2020). https://doi.org/10.11648/j.epes.20200904.12
4. S.M. Mousavi Gazafrudi, A. TabakhpourLangerudy, E.F. Fuchs, K. Al-Haddad, Power quality issues in railway electrification: a comprehensive perspective. IEEE Trans. Ind. Electron. **62**(5), 3081–3090 (2015). https://doi.org/10.1109/TIE.2014.2386794
5. W.-H. Ko, J.-C. Gu, Using a passive filter to suppress harmonic and resonance effects on railway power systems. J. Chin. Inst. Eng. **37**(7), 946–954 (2014).
6. V. Vasanthi, S. Ashok, Harmonic filter for electric traction system, in *2011 IEEE PES Innovative Smart Grid Technologies, India* (2011)
7. Z. Zhiwen, W. Bin, K. Jinsong, L. Longfu, A multipurpose balance Transformer for Railway Traction Applications. IEEE Trans. Power Deliv. **24**, 258–263 (2005)
8. V.Z. Manusov, U. Bumtsend, Y.V. Demin, Analysis of the power quality impact in power supply system of Urban railway passenger transportation—the city of Ulaanbaatar, in *2018 IOP Conference Series: Earth and Environmental Science*, vol. 177, pp. 012024 (2018)
9. F. Fathima, S. Prabhakar Karthikeyan, Harmonic analysis on various traction transformers in co-phase system. Ain Shams Eng. J. **7**, 627–638 (2016)
10. H. Myneni, S.K. Ganjikunta, S. Dharmavarapu, Power quality enhancement by hybrid DSTATCOM with improved performance in distribution system. Int. Trans. Electr. Energ. Syst. e12153 (2019). https://doi.org/10.1002/2050-7038.12153
11. I.S. Sujono, O.A. Qudsi, Application of D-STATCOM to reduce unbalanced load using synchronous reference frame theory, in *2020 10th Electrical Power, Electronics, Communications, Controls and Informatics Seminar (EECCIS)*, Malang, Indonesia, pp. 65–70 (2020). https://doi.org/10.1109/EECCIS49483.2020.9263476
12. T. Karthik, M. Prathyusha, R. Thirumalaivasan, M. Janaki, Power quality improvement using DSTATCOM, in *2019 Innovations in Power and Advanced Computing Technologies (i-PACT)*, Vellore, India, pp. 1–7 (2019). https://doi.org/10.1109/i-PACT44901.2019.8960234
13. B. Singh, A. Chandra, K. Al-Haddad, *Power Quality Problems and Mitigation Techniques* (Wiley, 2015)

Chapter 7
Mitigation of Harmonics Using Passive-Series Active-Hybrid Filter in 1-Φ and 3-Φ System Feeding Nonlinear Load

G. Jayachitra, H. A. Vidya, and Ruma Sinha

Abstract Extensive usage of power electronic systems in transmission and distribution systems causes unbalance, injects harmonics and deteriorates the power quality. A hybrid filter which is a combination of shunt passive and series active filter is proposed in this paper to eliminate the harmonic distortion in the current drawn from the source. The design and analysis are carried out in the MATLAB/SIMULINK environment, and the performance is analyzed. A comparative study is presented with passive filter, series active filter and hybrid filter.

Keywords Series active filter · Passive filter · Hybrid filter · Power quality

7.1 Introduction

Many loads used in industry and commercial applications, for example adjustable speed drive, power electronic converters, computer systems, air-conditioning systems, light dimmer etc., show nonlinear characteristics and cause power quality issues in the system [1]. With the growth of semiconductor technologies, power electronic devices are being very commonly used in electrical power systems for rectification, inversion and control of output power. This injects a huge amount of harmonics into the electrical transmission and distribution network and contaminates the power [2, 3]. Many of these nonlinear loads individually have very little effect, but when many of them are connected to the same system or if one of them draws very high power, then that affects the whole system [4]. If the current drawn from the supply is not sinusoidal, then the other loads connected in the system also get affected. As the harmonic content increases, the heating of the transformer core increases which in turn causes increase in power loss and results in reduction of efficiency. Researchers have proposed active filters to eliminate higher magnitude harmonics of lower order; however, they are not considered as economic solutions.

G. Jayachitra (✉) · H. A. Vidya · R. Sinha
Global Academy of Technology, Bangalore, India

A. Khosla and M. Aggarwal (eds.), *Smart Structures in Energy Infrastructure*, Studies in Infrastructure and Control,
https://doi.org/10.1007/978-981-16-4744-4_7

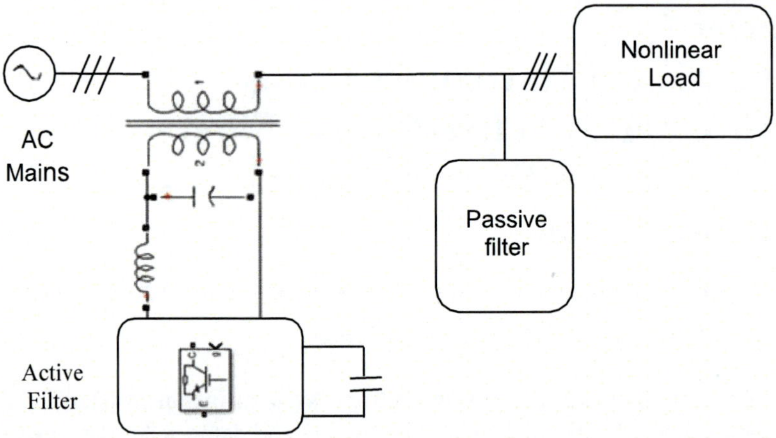

Fig. 7.1 Structure of hybrid filter

Passive filters are a low-cost solution, but they provide only fixed compensation and also may lead to resonance [2, 5].

This work proposes a hybrid structure with a voltage source converter (VSC)-based "series active filter" connected in series to the main supply and a passive filter connected across load. The harmonics generated by the "nonlinear load" flows through the series active filter and gets compensated thus maintaining the power drawn from supply harmonic free. The configuration is shown in Fig. 7.1. Hybrid filter uses a shunt-connected "passive filter" and a small-capacity "active filter." Harmonic distortion produced by the load is suppressed by the passive filter, while filtering attributes are improved by the passive filter.

7.2 Filter Design

7.2.1 Design of a Series "Active Filter"—Three Phase

"Series active filter" consists of a voltage source converter (VSC) with an inductor connected in series, a "coupling transformer" and a "DC bus capacitor," and is connected in series with the AC supply while feeding a nonlinear load [5–8]. Series active filter is employed to minimize the harmonic currents and to maintain almost unit power factor (UPF) at the AC supply side. The design of this filter includes calculation of the voltage, current, value of AC inductor and DC capacitor.

Voltage at the DC bus V_{DC} is dependent on the minimum voltage level required to obtain desired AC output.

Voltage across the bus capacitor is

$$V_{\mathrm{DC}} = \frac{2\sqrt{2}V_{\mathrm{LL}}}{\sqrt{3}m} \tag{7.1}$$

where m refers to the modulation index and V_{LL} is the RMS value of line voltage. Fundamental component of load AC voltage is given by

$$V_{\mathrm{LL}} = \left(\frac{\sqrt{6}}{\pi}\right) V_d \tag{7.2}$$

Series active filter voltage is calculated as

$$V_f = \sqrt{\left[\left(\frac{1}{\pi}\right)\left\{\int_{0}^{\frac{\pi}{3}}\left(V_{\mathrm{ph}}\sqrt{2}\sin\theta - \frac{V_d}{3}\right)^2 \mathrm{d}\theta + \int_{\frac{\pi}{3}}^{\frac{2\pi}{3}}\left(V_{\mathrm{ph}}\sqrt{2}\sin\theta - 2\frac{V_d}{3}\right)^2 \mathrm{d}\theta + \int_{\frac{2\pi}{3}}^{\pi}\left(V_{\mathrm{ph}}\sqrt{2}\sin\theta - \frac{V_d}{3}\right)^2 \mathrm{d}\theta\right\}\right]} \tag{7.3}$$

Calculation of Rated value of Current

Fundamental component of the current flowing through the load decides the current rating of the series "active power filter." Load power is calculated as

$$P_{\mathrm{DC}} = \frac{V_d^2}{R} \tag{7.4}$$

For a unity power factor (UPF) supply current and a neglecting losses in the series "active filter," the RMS value of the supply current is given by

$$I_f = \frac{P}{\sqrt{3}V_{\mathrm{LL}}} \tag{7.5}$$

where P is the input power.

Design of DC Capacitance of a "Series Active Filter"

"DC bus capacitance" can be calculated using the following expression.

$$(0.5)C_{\mathrm{DC}}(V_{\mathrm{DC}}^2 - V_{\mathrm{DC1}}^2) = 3V_f I_f \Delta t \tag{7.6}$$

where V_{DC} is the rated voltage; V_{DC1} is the permissible voltage drop in DC bus during transient condition; Δt is the time and C_{DC} is the "DC bus capacitance."

Design of an Interfacing Inductor of a Series Active Filter.

The value of the interfacing inductor depends on ripples present in current. Assuming 5% ripples and overloading factor "a" of 1.2, the inductance is obtained from

$$L_f = \left(\frac{N_{\mathrm{VSI}}}{N_{\mathrm{supply}}}\right)(\sqrt{3}/2)m_a V_{\mathrm{DC}}/(6af_s\Delta I_f) \tag{7.7}$$

Design of Ripple Filter

Design of ripple filter is carried out to eliminate the "switching frequency ripples" from the injected voltage of the series "active power filter." The ripple filter is tuned at 50% of the switching frequency f_r and is calculated for both single and three phases.

$$f_r = \frac{1}{2\pi R_f C_f} \tag{7.8}$$

Design of Series Active Filter—Single phase

"DC bus voltage" across the capacitor is

$$V_{\mathrm{DC}} = \frac{\sqrt{2}V_{\mathrm{SAF}}}{m} \tag{7.9}$$

where m is the modulation index and V_{SAF} is voltage across the series active filter.

Current rating

$$I_f = \frac{P}{V_s} \tag{7.10}$$

where P is the power drawn by the load and V_s is the supply voltage.

DC Capacitance

Capacitance for a single-phase series active filter can be calculated as

$$(0.5)C_{\mathrm{DC}}(V_{\mathrm{DC}}^2 - V_{\mathrm{DC1}}^2) = V_{\mathrm{SAF}} I_f \Delta t \tag{7.11}$$

where V_{DC} indicates the rated value of voltage; V_{DC1} indicates allowable drop in "DC bus voltage" during transients; Δt is the time; and C_{DC} is the "DC bus capacitance."

Interfacing Inductor

The magnitude of the interfacing inductor is dependent on ripples present in current.

Assuming 5% ripples, the inductor is calculated as

$$L_f = \frac{V_{\mathrm{DC}}}{4 f_s \Delta I_f} \tag{7.12}$$

where f_s is the switching frequency.

7.2.2 *Design of a Passive Filter—For Both Single and Three Phases*

Passive filter is a combination of inductor and capacitor connected to eliminate selected harmonic component. Size of the capacitor is computed based on the reactive power need Q_C of the load and is given as

$$C_n = \frac{Q_c}{m\omega V_s^2} \tag{7.13}$$

The inductance required to be connected for the "*n*th order filter" is obtained as

$$L_n = \frac{1}{n^2\omega^2 C_n} \tag{7.14}$$

The resistance to be connected with the inductor of the "*n*th order filter" is obtained as

$$R_n = \frac{n\omega L_n}{Q_n} \tag{7.15}$$

where Q_n is the "quality factor" of the inductor of the "*n*th order filter." Filter parameters C_H, L_H and R_H are calculated using Eqs. (7.13)–(7.15). From Eqs. (7.1) to (7.15), the parameters for series "active filter" and passive "shunt filter" are calculated and shown in Table 7.1.

Table 7.1 Designed values of filter parameters

	Series "active filter" parameters	Passive shunt filter parameters		
Single-phase system supplying nonlinear load	$C_f = 3.2$ μF, $L_f = 0.714$ mH $C_{\mathrm{DC}} = 406.52$ μF	$C_H = 59.41$ μF, $L_H = 18.96$ mH $R_H = 3.57$ Ω		
Three-phase system supplying nonlinear load	$C_f = 3.2$ μF, $L_f = 6.3$ mH $C_{\mathrm{DC}} = 33.04$ μF	$R_5 = 0.123$ Ω, $L_5 = 1.565$ mH, $C_5 = 259.21$ μF	$R_7 = 0.088$ Ω $L_7 = 0.798$ mH $C_7 = 259.21$ μF	$R_H = 0.558$ Ω $L_H = 0.323$ mH $C_H = 259.21$ μF

7.3 Simulation and Results

Detailed simulation study has been performed using "MATLAB SIMULINK" environment for single-phase and three-phase system supplying nonlinear load considering the cases without filter, with passive filter, with series active filter and a series active-shunt passive hybrid filter. Controlled rectifiers connected to R-L load have been used as nonlinear load. Simulink model of single-phase system with hybrid filter is shown in Fig. 7.2, and the waveforms are presented in Fig. 7.3.

For simulation, the supply voltage considered is 230 V at 50 Hz. Simulation study shows that without filter the total harmonic distortion (THD) in source current was 18.95, which was brought down to 5.05% with series active filter and to 0.4% with the hybrid filter. Harmonic analysis of input current is shown in Fig. 7.4.

Figure 7.5 shows the Simulink model of active series and passive shunt-based hybrid power filter with 3-Φ system supplying nonlinear load, while Fig. 7.6 shows the source side current and voltage waveforms for the same. System is tested without filter, with passive filter, with series active filter and with hybrid filter.

The FFT analysis shows that the input current as well as the input voltage waveforms is contaminated with harmonics. The shunt passive filter reduces the harmonic distortion but does not meet the standards as per IEEE—519. Series active filters bring the THD within specified limits, but hybrid filters bring down the THD to a very low value. THD in source side voltage and current have been shown in Tables 7.2 and 7.3, respectively, for the cases without filter and with different types of filters connected.

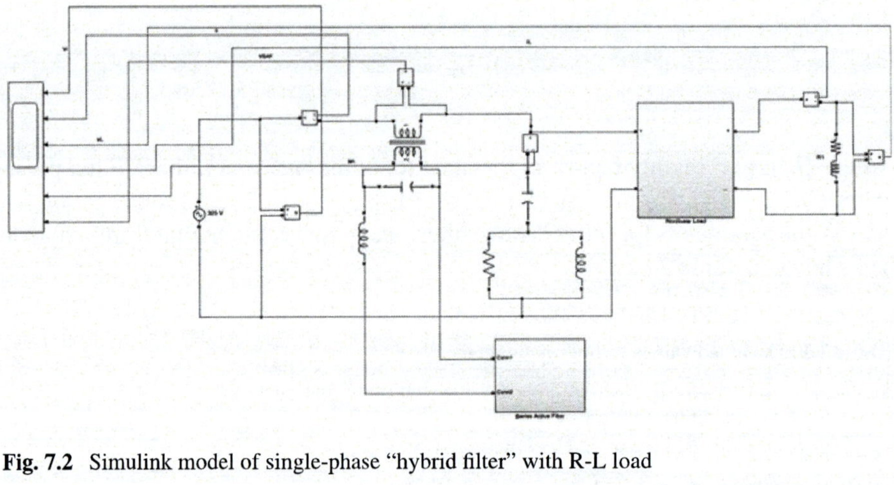

Fig. 7.2 Simulink model of single-phase "hybrid filter" with R-L load

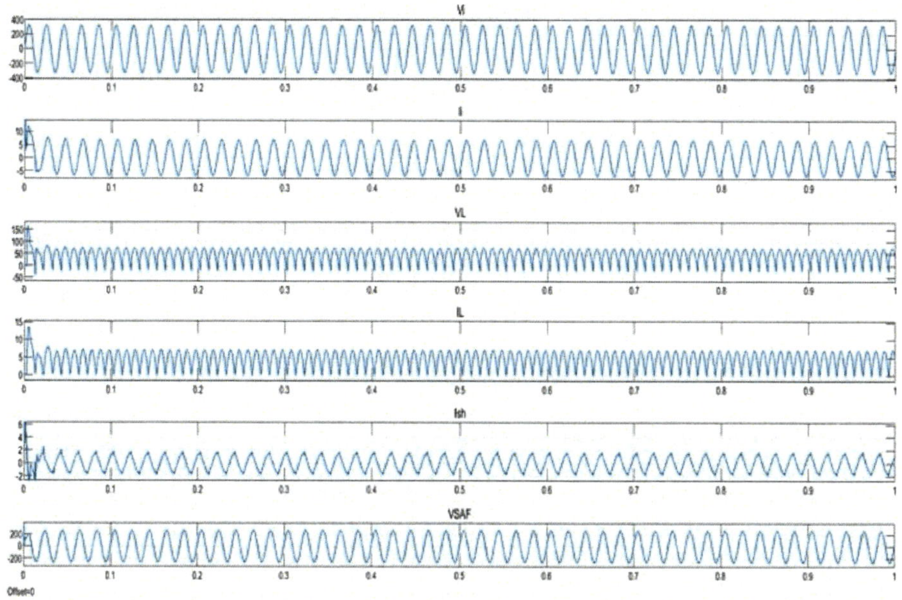

Fig. 7.3 Waveform with hybrid filter connected to single-phase system—source side voltage, output current, load voltage, load current, current through shunt passive filter and voltage across series active filter

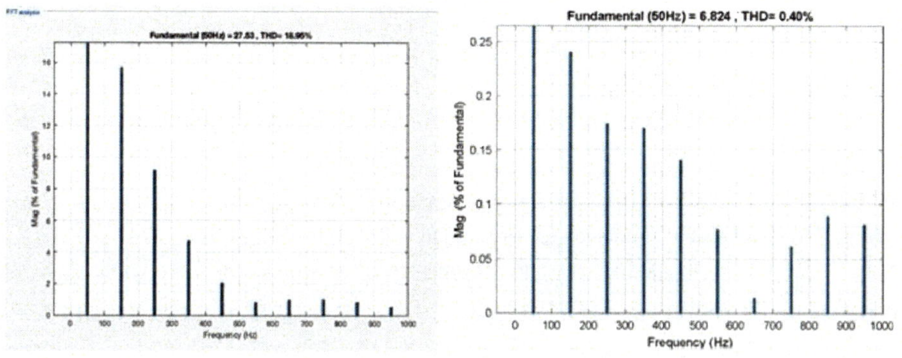

(a) (b)

Fig. 7.4 FFT analysis of input current waveform of single-phase system **a** without filter **b** with hybrid filter

7.4 Conclusion

A hybrid filter is presented which uses a combination of a series active and a passive shunt filter applied to single-phase and three-phase systems connected to nonlinear load. Simulation results and analysis show that the nonlinear load injects harmonic in the system, while the harmonic distortion can be reduced by the usage of series active

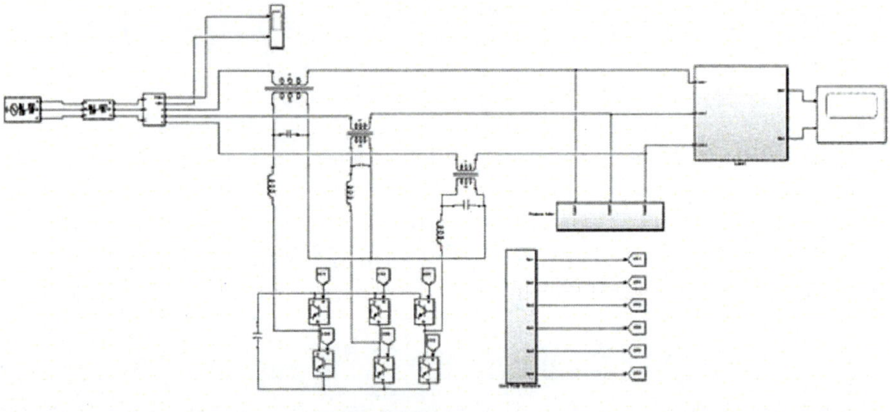

Fig. 7.5 Simulink model of three-phase hybrid filter

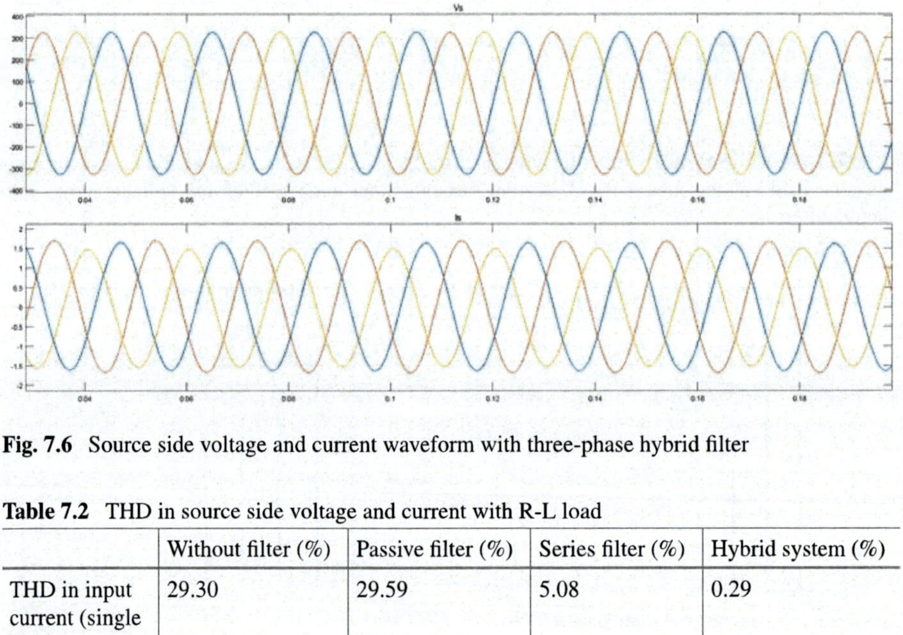

Fig. 7.6 Source side voltage and current waveform with three-phase hybrid filter

Table 7.2 THD in source side voltage and current with R-L load

	Without filter (%)	Passive filter (%)	Series filter (%)	Hybrid system (%)
THD in input current (single phase)	29.30	29.59	5.08	0.29
THD in input current (three phase)	6.74	2.02	2.01	0.26
THD in input voltage (three phase)	41.70	8.35	0.65	0.09

Table 7.3 THD in source side voltage and current with resistive load

	Without filter (%)	Passive filter (%)	Series filter (%)	Hybrid system (%)
THD in input current (single phase)	18.95	20.42	5.05	0.42
THD in input current (three phase)	41.14	5.90	0.63	0.08
THD in input voltage (three phase)	6.84	1.35	2.06	0.25

filter and almost eliminated by the usage of hybrid filter. Input current harmonics can be brought to 0.08% with the proposed hybrid filter from a level of 41.14% when no filter is connected.

References

1. H.A. Kazem, Harmonic mitigation techniques applied to power distribution networks. Adv. Power Electron. **2013**, article ID 591680
2. B. Singh, K. Al-Haddad, A. Chandra, Harmonic elimination, reactive power compensation and load balancing in three-phase, four-wire electric distribution systems supplying non-linear loads. Electr. Power Syst. Res. **44**(2), 93–100 (1998). ISSN 0378-7796
3. J. Pinzón, A. Pedraza, F. Santamaría, Harmonic impact of non-linear loads in a power distribution system, in *Conferencia Brasileña sobre Calidad de Energía Eléctrica (CBQEE)*, Brasil, July 2013, vol. 10
4. P.K. Joshi, S.S. Bohra, Simulation of single-phase shunt active power filter for domestic non-linear loads. 978-1-4673-6322-8/13/$31.00_c 2013 IEEE
5. B. Singh, A. Chandra, K. Al-Haddad, *Power Quality: Problems and Mitigation Techniques* (Wiley, 2014)
6. P. Kishore Kumar, M. Sharanya, Design of hybrid series active filters for harmonic reduction in single phase systems. Int. J. Mod. Eng. Res. (IJMER) **3**(5) (2013)
7. A.P. Bagde, R.B. Ambatkar, R.G. Bhure, B.S. Rakhonde, Power quality improvement by series active power filter—a review. Int. Res. J. Eng. Technol. (IRJET) **04**(01) (2017)
8. V. Nakade, S. Patil, Implementation if power quality enhancement using hybrid series active filter, in *2019 International Conference on Communication and electronic Systems (ICCES)*, Coimbatore, India, 2019, pp. 238–241

Chapter 8
Thermal Modelling of Solar Photovoltaic Panel Using FEM

Abhilash Narasimhan

Abstract The significance of solar energy is increasing in the modern world. Therefore, it is imperative to harness this energy to its maximum potential. The temperature of the solar panel increases due to extended exposure to sunlight. The objective of this study is to analyze the thermal performance of a solar panel. The solar panel heat flow is analyzed in the finite element approach, by modelling the different layers in the solar panel. The heat flow equations and the boundary conditions for this analysis are discussed in this paper. The results are simulated with various conditions of varying ambient temperature and solar fluxes. The three-dimensional temperature profile is shown, and this methodology can be used for optimized material selection for solar panel.

Keywords Temperature profile · FEM · Solar panel

8.1 Introduction

Solar energy is a fast developing industry that is abundant and renewable form of energy. India is also a country with high-national potential because of the presence of approximately 280–300 sunny days in most parts of the country. Because of such a high potential, the photovoltaic industry is increasing and rapidly making progress. For example, India's solar capacity is 36.9 GW at the end of 2020 [1], and it has superseded the target of 20 GW by 2022. Therefore, the need to increase the efficiency of the solar energy conversion becomes primal [2–4]. Solar energy is converted to electrical energy when there is incidence of photons on the photovoltaic cell transfers the energy from the photons to the charge carriers. An electric field across the junctions helps creates the current to flow across a load that is connected. There are several advantages if the efficiency of the photovoltaic conversion increases some of them being the reduction in the surface area of PV plant, reduction in production costs, etc. The solar cell temperature is one of the critical factors that affect the PV

A. Narasimhan (✉)
Swelect Energy Systems, Chennai, India

A. Khosla and M. Aggarwal (eds.), *Smart Structures in Energy Infrastructure*,
Studies in Infrastructure and Control,
https://doi.org/10.1007/978-981-16-4744-4_8

module efficiency [5, 6]. The temperature of the PV cell affects the performance of the solar cell [7–9]. As the temperature of the photovoltaic (PV) cell increases, the efficiency of solar cell reduces [10–12]. Several attempts are being made world over to optimize the temperature of operation for the solar cells like that in [13–15].

In this work, a study has been carried out where PV panel has been considered where the geometry is divided into the different layers, and the thermal analysis has been carried out for the same. This approach helps in numerically simulating in computer simulation softwares. The study that has been carried out here uses three-dimensional model for predicting the behaviour of the PV panel in thermal aspect.

8.2 Theoretical Background and Methodology

The PV module consists primarily of three layers. The layer that faces the sun directly or the atmosphere is the glass layer that generally has anti-reflection coatings on top to ensure that maximum incident energy is absorbed. The second layer contains the solar cells lined up in series and parallel connections. This layer is responsible for converting the solar energy to electrical energy. The third layer is called as the back sheet. The back sheet is an important component of the PV panel as this provided strength to the panel, protection from moisture and also electrical insulation. The durability of the back sheet plays a major role in longevity of the PV panels. The back sheet of the PV panel consists of fluoropolymers. In this study, the back sheet that is employed for study is Kynar polyvinylidene fluoride (PVDF) films.

In the numerical analysis that has been carried out, all the materials are assumed to be isotropic and they do not have any temperature dependence. The convection coefficient at the glass surface is assumed to be half of the back sheet surface. The heat flow across all the layers and the distribution of the same needs to be understood. In order to predict the temperatures, the followings are the equations considered. The solar flux is given by G in the following equations.

$$\rho c \frac{\partial T}{\partial t} = \nabla.(\lambda \nabla T) \tag{8.1}$$

where (8.1) is the heat transfer equation, where λ is a tensor. The objective is to solve for the temperature at any point in a PV panel which is given by T (x, y, z). The evaluation is to be performed for all the three layers. The initial condition of the system is given by T $(x, y, z) = T_0$. The boundary conditions for glass, solar cells and kynar are given in Eqs. (8.2), (8.3), and (8.4), respectively, to solve in finite element method.

$$\int \rho_g c_g T_g \left(\frac{\partial T_g}{\partial t}\right) \mathrm{d}V - \int T_g \alpha_g \tau_g G \mathrm{d}S$$

$$-\int T_g\left(\alpha_g G - h_{g-a}\left(T_g - T_a\right) - \sigma\varepsilon\left(T_g^4 - T^4\right)\right)\mathrm{d}S$$
$$+\int \nabla T_g\left(\lambda_g \nabla T_g\right)\mathrm{d}V = 0 \tag{8.2}$$

$$\int \rho_c c_c T_c\left(\frac{\partial T_c}{\partial t}\right)\mathrm{d}V - \int T_c\alpha_c\tau_g G\mathrm{d}S + \int \nabla T_c(\lambda_c \nabla T_c)\mathrm{d}V$$
$$+\int T_c(\eta_{\mathrm{ref}}\beta_0 T_c G)\mathrm{d}V - \int T_c\eta_{\mathrm{ref}}G(1+\beta_0 T_{\mathrm{ref}})\mathrm{d}V = 0 \tag{8.3}$$

$$\int \rho_{\mathrm{kyn}}c_{\mathrm{kyn}}T_{\mathrm{kyn}}\left(\frac{\partial T_{\mathrm{kyn}}}{\partial t}\right)\mathrm{d}V + \int \nabla T_{\mathrm{kyn}}\left(\lambda_{\mathrm{kyn}}\nabla T_{\mathrm{kyn}}\right)\mathrm{d}V$$
$$+\int T_{\mathrm{kyn}}\left(-h_{\mathrm{kyn}-a}\left(T_{\mathrm{kyn}} - T_a\right) - \sigma\varepsilon_{\mathrm{kyn}}\left(T_{\mathrm{kyn}}^4 - T^4\right)\right)\mathrm{d}S = 0 \tag{8.4}$$

The elemental equations for each layer consists of finite elements that are connected at nodes. The geometry of these elements is hexahedral. The simplified equations for Eqs. (8.2)–(8.4). The simplified elemental equation in a layer is given as

$$[c_{\mathrm{layer}}^e]\dot{T}_{\mathrm{layer}}^e + \left[k_{\mathrm{layer}}^e\right]T_{\mathrm{layer}}^e = f_{\mathrm{layer}}^e \tag{8.5}$$

where $[c^e]$ is the heat capacity matrix, $[k^e]$ is the thermal conductivity matrix and T^e and f^e are the nodal vector temperature and nodal flow of element, respectively. These elementary equations for each layer are combined into global assembly equations of the form

$$[C]\dot{T} + [K]T = F \tag{8.6}$$

The above matrix equations are solved by employing Newton Raphson methodology.

The parameters that are considered for all the layers are shown in Table 8.1.

Table 8.1 Thermal parameters

Parameter	Glass	SolarCell	Kynar
Thermal conductivity	1.75	1.5	0.2
Density	3500	2350	1250
Specific heat capacity	550	650	1200
Transmissivity	0.9	0.02	0.013
Emissivity	0.9	–	0.9
Absorptivity	0.05	0.9	0.13
Reflectivity	0.05	0.09	0.9

8.3 Results

The meshing of the solar panel is as shown in Fig. 8.1. For the solar flux of 1000 W/m^2 and ambient temperature of 50 °C, the solution of the global assembly equations gives the temperature profile as shown in Fig. 8.2.

The temperature of the PV cell is measured when the ambient temperature is 25 °C, and the solar intensity is varied from 200 to 1000 W/m^2 as shown in Fig. 8.3.

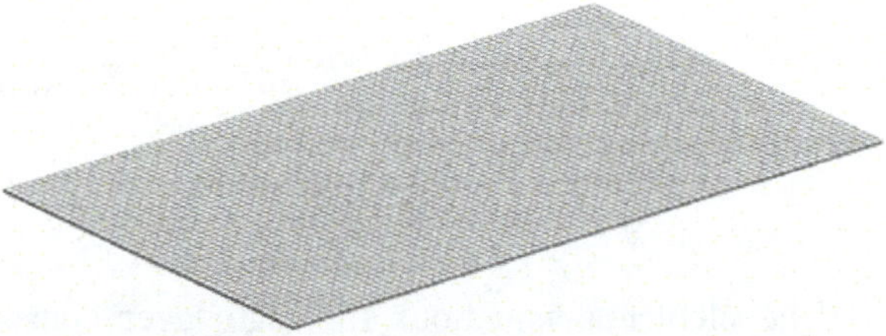

Fig. 8.1 Finite element mesh

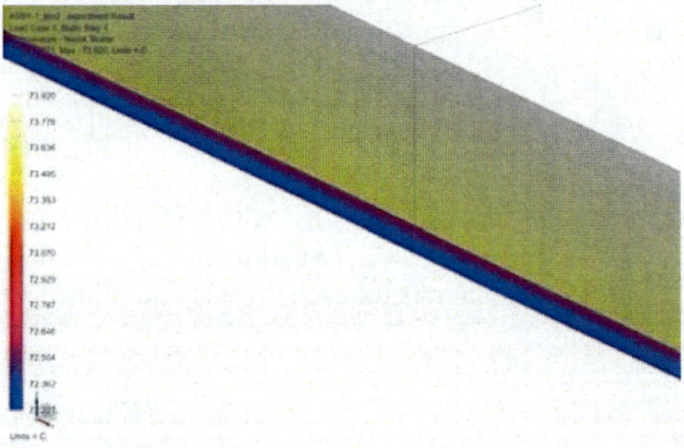

Fig. 8.2 Three-dimensional profile of a solar panel

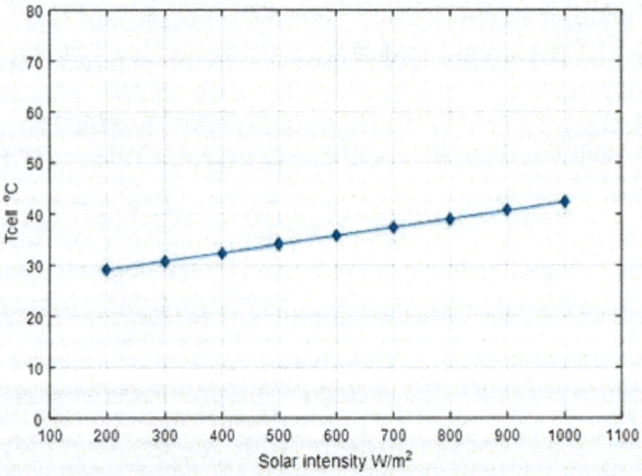

Fig. 8.3 Temperature of PV cell when ambient temperature is 25 °C

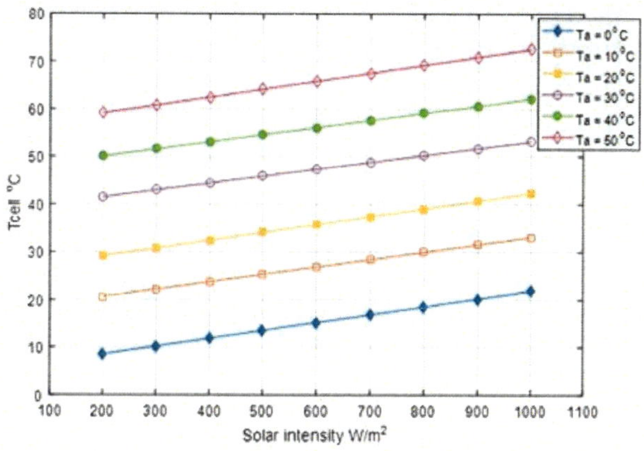

Fig. 8.4 The variation of temperature of PV cell with solar intensity when ambient temperature is varied

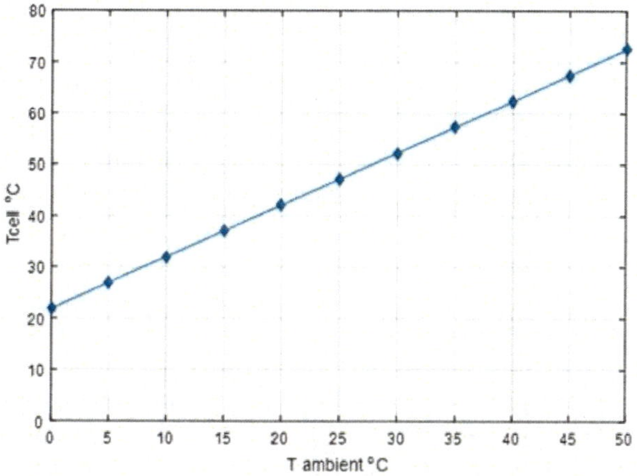

Fig. 8.5 PV cell temperature variation with variant ambient temperature, when solar flux is constant at 1000 W/m^2

The variation of temperature of PV cell with solar intensity when ambient temperature is varied is as shown in Fig. 8.4. The individual PV cell variation with variant ambient temperature is as shown in Fig. 8.5. The solar flux is kept constant at 1000 W/m^2. Similarly, the variation of temperature of the PV cell with ambient temperature when the solar flux is varied is as shown in Fig. 8.6.

8.4 Conclusion

From the above results, it can be observed that the temperature of the cell increases with the increase in ambient temperature. Higher solar intensity even at lower ambient temperatures increases the cell temperature. Higher solar intensity in this case reduces the electrical efficiency, whilst the thermal efficiency increases. Also, the region of solar cell has higher temperature than the back sheet kynar. Therefore, cooling of the panel needs to be performed from the back side of the panel by having liquid

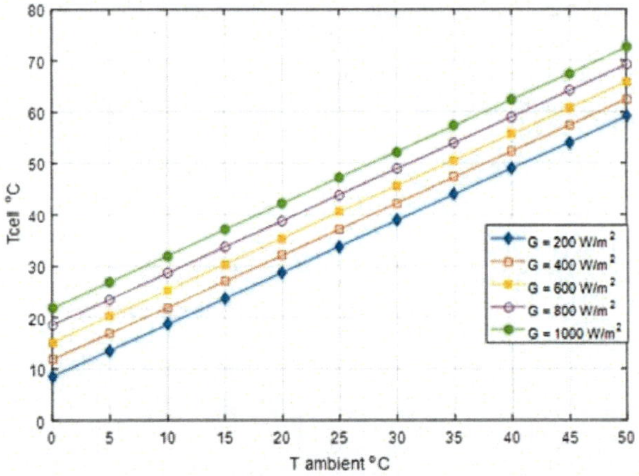

Fig. 8.6 Variation of temperature when solar flux is varied

coolants. At the front glass, air cooling can help. Natural cooling effect is obtained by higher wind speeds. Therefore, by having lesser ambient temperatures and cooling effects, higher performance of the solar panels can be achieved. Additionally, by varying the thermal parameters in the equations, an optimum material can be chosen for higher efficiency for better control in industry.

References

1. Progress Physical: Ministry of new & renewable energy. Programme/Scheme wise Physical Progress in 19 (2018)
2. I. Purohit, P. Purohit, Performance assessment of grid-interactive solar photovoltaic projects under India's national solar mission. Appl. Energy **222**, 25–41 (2018)
3. R.R. Kannan, G. Madhumita, Smart grid—introduction, advantages and implementation status in India with a focus on Tamil Nadu: a systematic review. Int. J. Adv. Sci. Technol. **29**(3s), 1146–1156 (2020)
4. MI Smart Grids, Smart Grids Innovations Challenge Country Report (2019)
5. D. Furkan, E.M. Mehmet, Critical factors that affecting efficiency of solar cells. Smart Grid Renew. Energy (2010)
6. H.J. Queisser, J. Werner, Principles and technology of photovoltaic energy conversion, in *Proceedings of 4th International Conference on Solid-State and IC Technology* (IEEE, 1999), pp 146–150
7. F. Brihmat, S. Mekhtoub, PV cell temperature/pv power output relationships homer methodology calculation, in *Conference Internationale des Energies Renouvelables CIER'13*/Int. J. Sci. Res. Eng. Technol. **1** (International Publisher & C. O, 2014)
8. S. Dubey, J.N. Sarvaiya, B. Seshadri, Temperature dependent photo-voltaic (pv) efficiency and its effect on pv production in the world—a review. Energy Procedia **33**, 311–321 (2013)
9. Q. Hassan, M. Jaszczur, E. Przenzak, J. Abdulateef, The PV cell temperature effect on the energy production and module efficiency. Contemp. Probl. Power Eng. Environ. Prot. **33** (2016)
10. H. Zondag, Flat-plate pv-thermal collectors and systems: a review. Renew. Sustain. Energy Rev. **12**(4), 891–959 (2008)
11. D.E. Jung, C. Lee, K.H. Kim, S.L. Do, Development of a predictive model for a photovoltaic module's surface temperature. Energies **13**(15), 4005 (2020)

12. V. Chayapathy, G. Anitha, P. Raghavendra, R. Vijaykumar, Solar panel temperature prediction by artificial neural networks, ın *Proceedings of the 4th International Conference on Recent Trends on Electronics, Information, Communication & Technology (RTEICT-2019)*, Bengaluru, India (2019), pp. 17–18
13. Y. Lee, A.A. Tay, Finite element thermal analysis of a solar photovoltaic module. Energy Procedia **15**, 413–420 (2012)
14. J. Prakash, Transient analysis of a photovoltaic-thermal solar collector for co-generation of electricity and hot air/water. Energy Convers. Manag. **35**(11), 967–972 (1994)
15. A. Tiwari, M. Sodha, A. Chandra, J. Joshi, Performance evaluation of photovoltaic thermal solar air collector for composite climate of India. Sol. Energy Mater. Sol. Cells **90**(2), 175–189 (2006)

Chapter 9
Selective Harmonic Elimination for Cascade Multilevel Inverter Using Genetic Algorithm

Hemant Gupta, Arvind Yadav, and Sanjay Maurya

Abstract Multilevel inverter is popular for its merits and good characteristics. Among many topologies of multilevel inverter, cascade topology is known for its modular construction and viability in high-power applications. The low frequency selective harmonic elimination technique comprising nonlinear transcendental equations which is used to reduce the harmonics of output voltage. There are various methods of optimization to solve nonlinear transcendental equations out of which genetic algorithm is the biological evolutionary mechanism. This paper work analyzes the operation of an eleven-level cascade multilevel inverter which uses the selective harmonic exclusion approach to evaluate the values of optimal switching angles using genetic algorithm in order to minimize the harmonics of the output voltage.

Keywords Multilevel inverter (MLI) · Cascade H-bridge (CHB) · Selective harmonic elimination (SHE) · Pulse width modulation (PWM) · Total harmonic distortion (THD) · Genetic algorithm (GA)

9.1 Introduction

In the recent trend, multilevel inverters are one of the most enticing inverters which is known for its superior qualities such as its comparatively more appropriate output voltage waveform, preferable harmonic performance, modularity in structure and relatively lower rating of semiconductor devices [1]. The increasing demand for multilevel inverter has contributed to the creation of different configurations. The key

H. Gupta (✉) · A. Yadav · S. Maurya
GLA University, Mathura, India
e-mail: hemant.gupta@gla.ac.in

A. Yadav
e-mail: arvind.yadav@gla.ac.in

S. Maurya
e-mail: sanjay.maurya@gla.ac.in

A. Khosla and M. Aggarwal (eds.), *Smart Structures in Energy Infrastructure*,
Studies in Infrastructure and Control,
https://doi.org/10.1007/978-981-16-4744-4_9

goal of the implementation of cascaded MLIs is to reach more voltage levels and thus improved power output, but with less components mainly semiconductor devices and related control circuits [2]. Essentially, there are three key multilevel inverter topologies that are well known in the industry. These topologies include: cascaded H-bridge topology of multilevel inverter (CHB-MLI), diode-clamped topology of multilevel inverter (DC-MLI) and flying capacitor topology of multilevel inverter (FC-MLI). The CHB-MLI is recommended for high voltage and high-power applications. This is since, of the three simple topologies, the CHB-MLI demands the lesser number of power electronics switches and units [3]. Moreover, it has a modular form that is robust relative to the DC-MLI and FC-MLI, and it can be built and operated on a large range of levels at a low degree of difficulty.

There have been several experiments being carried out over the past few years to eliminate harmonics in the switching converters [4]. The modulation pattern based on sinusoidal pulse width can be used to efficiently eliminate harmonics, but the switching frequency is high in this method, resulting in higher switching losses [5].

To keep the switching losses at minimum level, a low-frequency approach may be used [6]. One of the low-frequency tactics used here is SHE. Among the many optimum approaches, genetic algorithm is one of the most appropriate evolutionary approaches to solve the nonlinear SHE equations [7]. In order to refine the PWM sequence and to minimize the THD, SHE-PWM was developed in conjunction with a genetic algorithm artificial intelligence technique [8].

The CHB-based topology for eleven levels was introduced in this paper with the SHE technique added to it. Figure 9.1 shows a CHB-based topology with five cells to achieve eleven-level phase output voltage, in which five cells needed to get eleven levels of phase output voltage which is determined by the equation $N = 2M + 1$, where M is the number of cells needed and N indicates the number of levels demanded. Each CHB-MLI cell is comprised of a DC voltage source linked to one complete H-bridge as shown in Fig. 9.1 and a full H-bridge consisting of four-power

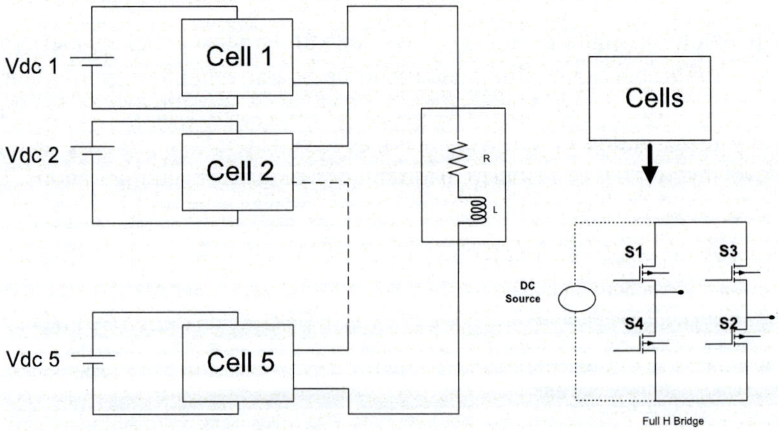

Fig. 9.1 Block diagram of 11-level MLI

semiconductor devices S1, S2, S3, and S4. MLI needs a separate DC source that is capable of generating the output voltage level in relation to the number of cells used in inverter unequal DC sources can be used in the cascade H-bridge inverter, known as asymmetrical configuration, to achieve higher levels without increasing the number of cells.

9.2 Calculations Included in SHE Technique

SHE is a technique of low-frequency control method that allows a power switch to be switched just one or two times within a one cycle and has low value of switching losses and electromagnetic interference, whereas the other techniques of sinusoidal PWM are used as a form of high-frequency switching methods [9]. In addition, the dominant low-order harmonics can be removed by the SHE method, causing the filter size at the inverter output to be minimized. Using Fourier's expansion, the output-phase voltage waveform can be represented as the sum of periodic sine and cosine signals and a constant. Due to the waveform's quarter symmetry, cancels harmonics and DC constant [10]. The output voltage's waveform can therefore be written as

$$V_n = \sum_{3,5,7,9}^{\infty} \frac{4Vdc}{n\pi}(\cos(n\alpha_1) + \cos(n\alpha_2) + \cdots + \cos(n\alpha_s))\sin(n\omega t) \tag{9.1}$$

where V_n is the magnitude of the nth harmonic voltage and S denotes the number of cells in the multilevel inverter. To exclude a certain number of harmonics, a certain number of switching angles must be calculated using the SHE equation [11]. For each inverter configuration, a collection of algebraic equations is formulated that determines the number of switching angles. Figure 9.1 shows a single-phase 11-level inverter with five optimum switching angles to be determined based on a given series of nonlinear transcendental equations in order to eliminate four distinct harmonics.

$$\cos(\alpha_1) + \cos(\alpha_2) + \cos(\alpha_3) + \cos(\alpha_4) + \cos(\alpha_5) = 4m_i \tag{9.2}$$

where

$$m_i = \frac{\pi V1}{4SVdc} \tag{9.3}$$

$$\begin{aligned}
&\cos(3\alpha_1) + \cos(3\alpha_2) + \cos(3\alpha_3) + \cos(3\alpha_4) + \cos(3\alpha_5) = 0 \\
&\cos(5\alpha_1) + \cos(5\alpha_2) + \cos(5\alpha_3) + \cos(5\alpha_4) + \cos(5\alpha_5) = 0 \\
&\cos(7\alpha_1) + \cos(7\alpha_2) + \cos(7\alpha_3) + \cos(7\alpha_4) + \cos(7\alpha_5) = 0f(i) \\
&\cos(9\alpha_1) + \cos(9\alpha_2) + \cos(9\alpha_3) + \cos(9\alpha_4) + \cos(9\alpha_5) = 0
\end{aligned} \tag{9.4}$$

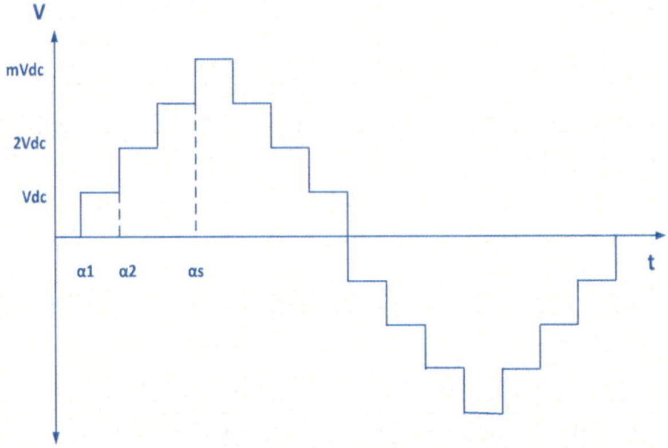

Fig. 9.2 Stepped output voltage waveform for m-level inverter

Figure 9.2 shows that all switching angles are less than 90° and are arranged in ascending order as follows:

$$\alpha_1 < \alpha_2 < \alpha_3 < \cdots < \alpha_s < 90^\circ \tag{9.5}$$

9.3 Essentials of Genetic Algorithms

A genetic algorithm is a computer model that solves problems of optimization by imitating genetic processes and evolutionary theory [12]. GA is a search tool that resembles biological evolutionary processes to identify the optimum value of the given problem [13]. Usually, this genetic algorithm is used to achieve a nearly global optimal solution. Each GA iteration is a new collection of strings known as chromosomes with enhanced fitness function, generated using genetic operators [14]. The GA is simple approach for any number of levels in multilevel inverter, and it can be extended to any number of levels; it does not require comprehensive derivations and analytical expressions for both removing and minimizing harmonics [15]. There are so many benefits of genetic algorithms, such as

- Gives just one solution for one index of modulation.
- There is no initial guess needed.
- It is not time intensive.
- It can be extended on any number of levels.

The main advantages of this approach are that it begins looking arbitrarily and manages vast quantities of data simultaneously.

9.3.1 Evaluation of the First Random Population

The first step of this algorithm begins with an initial random population being generated. The population can range from a few individuals to thousands. Each constituent of the population is measured after the individual's development, and the fitness of each individual is determined. The health benefit indicates how fit the member is in our optimal scenario.

9.3.2 Selection

After the appraisal, in order to boost the overall fitness, we will reject the poor constituents and only maintain the better people. Different selection strategies exist, but the basic principle is to pick a fitter individual for the next generation.

9.3.3 Crossover

In this step, by associating facets of our chosen members, we build new members. The idea is to merge the traits of two or more members to create a fitter infant that has the best features of its parents.

9.3.4 Mutation

The off-springs of mutation are created by modifying the genes of the member parents spontaneously. The algorithm appends the random vector to the parent for unconstrained problems. We start from the appraisal method again before we get a termination state (Fig. 9.3).

By using the GA codes in the environment of MATLAB, the nonlinear equations defined in Eq. (9.4) were resolved in order to fulfill the following objective function:

$$\text{Objective Function} = \sum_{i=3}^{i\text{th}} f(i) + \text{THD} \tag{9.6}$$

where

$$\text{THD} = \sqrt{\frac{\sum_{i=3}^{i\text{th}} i\text{th harmonic RMS voltage}}{\text{fundamental frequency RMS voltage}}} \tag{9.7}$$

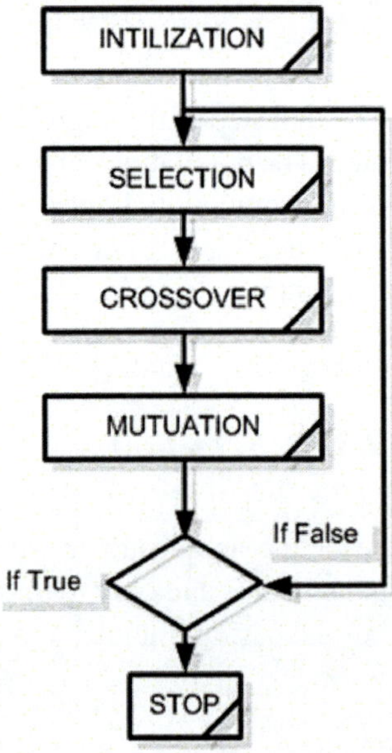

Fig. 9.3 Flow chart of genetic algorithm

9.4 Simulation Result

This section of this paper shows the values of optimal switching angles for an 11-level multilevel inverter as given in Table 9.1 whose values are used to obtained by genetic algorithm. The output voltage waveform of 11-level multilevel inverter is shown in Fig. 9.4, and the waveform of line voltage and line current at 0.8 power factor is also shown in Fig. 9.5. The THD spectra (7.10%) of line voltage for 11-level multilevel inverters are seen in Fig. 9.6. Five nonlinear equations are developed in this paper by the SHE method to remove four particular harmonics as 3rd, 5th, 7th, and 9th.

Table 9.1 Switching angles and THD

S. No.	MI	Switching angles (in degrees)					THD (in %)	
		θ_1	θ_2	θ_3	θ_4	θ_5	$1-\phi$	$3-\phi$
1	**0.2**	8.14	21.28	36.22	55.12	66.04	**12.32**	**7.15**
2	**0.3**	10.49	23.67	38.87	54.48	67.87	**13.72**	**7.10**
3	**0.4**	8.81	23.38	38.58	55.81	68.32	**13.38**	**8.16**
4	**0.5**	10.46	21.33	39.60	55.38	68.46	**13.68**	**8.21**
5	**0.6**	8.28	22.66	36.37	56.74	68.45	**13.08**	**8.77**
6	**0.7**	8.41	22.59	39.09	56.54	58.39	**15.29**	**8.69**
7	**0.8**	8.35	23.63	39.61	55.94	69.72	**14.05**	**8.13**
8	**0.9**	9.73	22.95	38.80	56.93	67.02	**14.25**	**7.7**
9	**1**	8.65	21.24	36.19	55.93	67.83	**12.45**	**8.15**

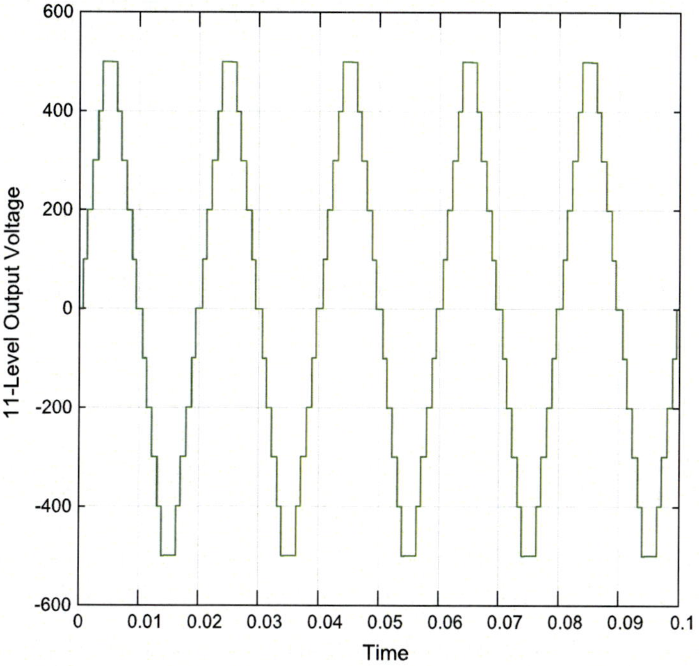

Fig. 9.4 Output phase voltage waveform of 11-level MLI

The graph between the optimum values of the switching angles and the modulation index is seen in Fig. 9.7.

9.5 Conclusion

In this paper, optimum switching angles for 11-level MLI were determined using a genetic algorithm technique. The aim of determining the optimum switching angles is to keep the value of THD as low as possible. The chosen harmonics are eliminated in this paper using SHE and THD got reduced. At the same time, simulation results have been verified with optimum switching angles obtained by the genetic algorithm.

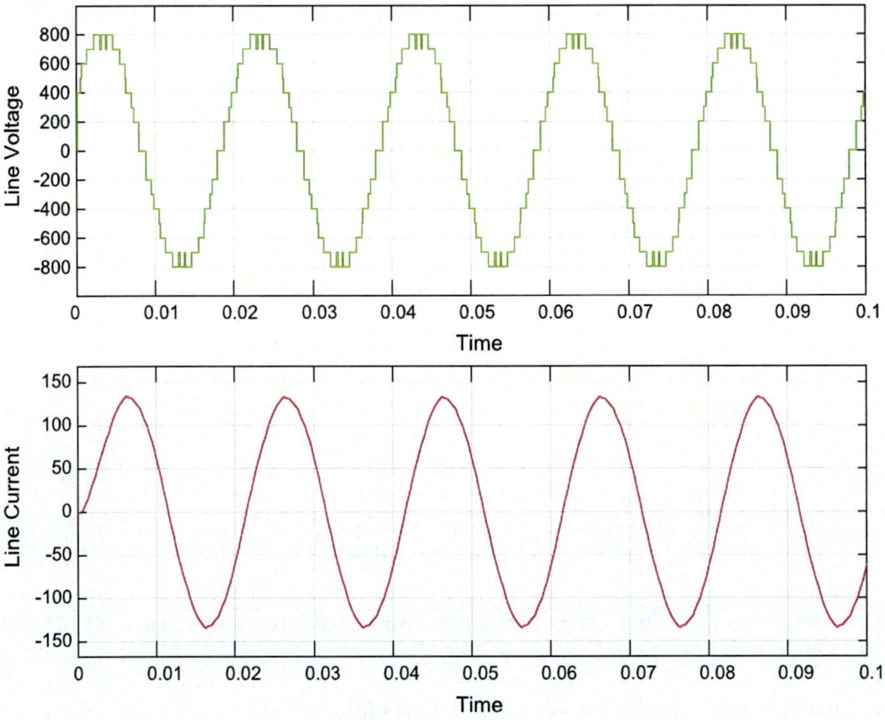

Fig. 9.5 Output line voltage and line current waveform of 11-level MLI

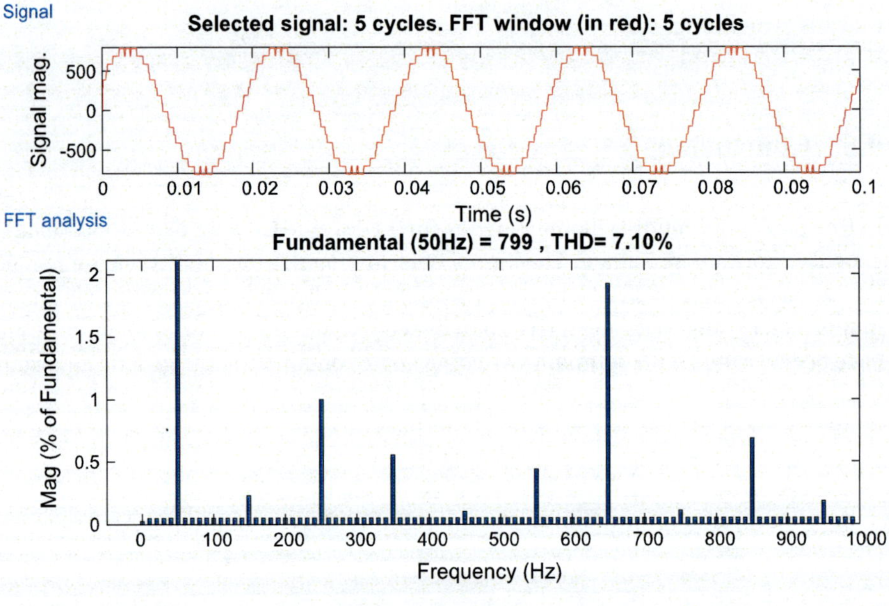

Fig. 9.6 THD spectra of line voltage for 11-level MLI

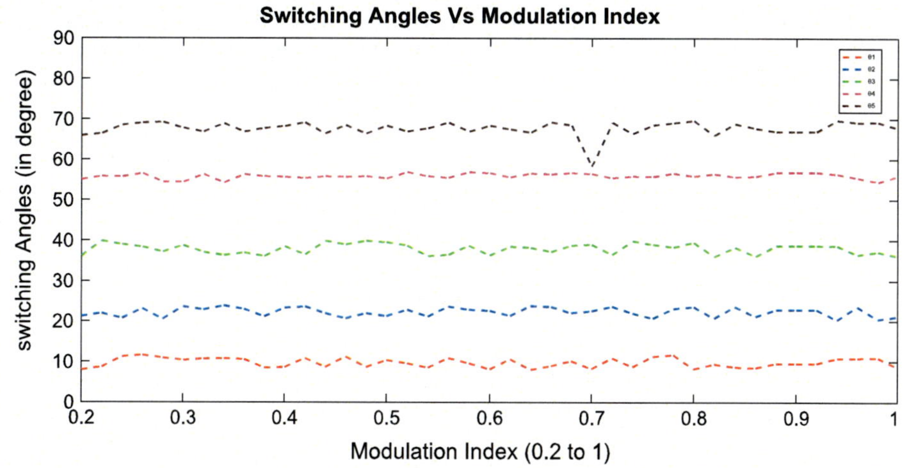

Fig. 9.7 Switching angles versus modulation index

References

1. M. Srndovic, A. Zhetessov, T. Alizadeh, Y.L. Familiant, G. Grandi, A. Ruderman, Simultaneous selective harmonic elimination and THD minimization for a single-phase multilevel inverter with staircase modulation. IEEE Trans. Ind. Appl. **54**(2), 1532–1541 (2018). https://doi.org/10.1109/TIA.2017.2775178
2. B. Ahmed, K.A. Aganah, M. Ndoye, M.A. Arif, C. Luciano, G.V. Murphy, Single-phase cascaded multilevel inverter topology for distributed DC sources, in *2017 IEEE 8th Annual Ubiquitous Computing, Electronics and Mobile Communication Conference (UEMCON)*, New York, NY, 2017, pp. 514–519. https://doi.org/10.1109/UEMCON.2017.8248980
3. H. Gupta, A. Yadav, S. Maurya, Cascade multilevel inverter based topology with reduced switch count, in *2020 4th International Conference on Electronics, Communication and Aerospace Technology (ICECA)*, Coimbatore, India, 2020, pp. 150–154. https://doi.org/10.1109/ICECA49313.2020.9297539
4. H.R. Massrur, T. Niknam, M. Mardaneh, A.H. Rajaei, Harmonic elimination in multilevel inverters under unbalanced voltages and switching deviation using a new stochastic strategy. IEEE Trans. Ind. Inf. **12**(2), 716–725 (2016). https://doi.org/10.1109/TII.2016.2529589
5. C. Buccella, M.G. Cimoroni, M. Tinari, C. Cecati, A new pulse active width modulation for multilevel converters. IEEE Trans. Power Electron. **34**(8), 7221–7229 (2019). https://doi.org/10.1109/TPEL.2018.2878967
6. H. Gupta, A. Yadav, S. Maurya, Multi carrier PWM for cascade topology of multilevel inverter, in *2020 International Conference on Recent Trends on Electronics, Information, Communication & Technology (RTEICT)*, Bangalore, India, 2020, pp. 328–332. https://doi.org/10.1109/RTEICT49044.2020.9315586
7. D. Chen, Y. Liu, J. Zhou, Optimized neural network by genetic algorithm and its application in fault diagnosis of three-level inverter, in *2019 CAA Symposium on Fault Detection, Supervision and Safety for Technical Processes (SAFEPROCESS)*, Xiamen, China, 2019, pp. 116–120. https://doi.org/10.1109/SAFEPROCESS45799.2019.9213395
8. A.A.K. Arani, A. Ghasemi, H. Karami, M. Akhbari, G.B. Gharehpetian, Optimal switching algorithm for different topologies of 15-level inverter using genetic algorithm, in *2019 5th Conference on Knowledge Based Engineering and Innovation (KBEI)*, Tehran, Iran, 2019, pp. 352–358. https://doi.org/10.1109/KBEI.2019.8734966

9. H. Gupta, A. Yadav, S. Maurya, Multi carrier PWM and selective harmonic elimination technique for cascade multilevel inverter, in *2016 2nd International Conference on Advances in Electrical, Electronics, Information, Communication and Bio-Informatics (AEEICB)*, Chennai, India, 2016, pp. 98–102. https://doi.org/10.1109/AEEICB.2016.7538405
10. H. Hashim, I. Abdel-Qader, Generalized solutions for THD investigation in cascaded multicell multilevel converters modulated with selective harmonic elimination PWM switching method, in *2016 IEEE 7th Annual Ubiquitous Computing, Electronics & Mobile Communication Conference (UEMCON)*, New York, NY, USA, 2016, pp. 1–7. https://doi.org/10.1109/UEMCON.2016.7777875
11. A. Moeini, S. Wang, Asymmetric selective harmonic elimination technique using partial derivative for cascaded modular active rectifiers tied to a power grid with voltage harmonics, in *2016 Asia-Pacific International Symposium on Electromagnetic Compatibility (APEMC)*, Shenzhen, China, 2016, pp. 982–987. https://doi.org/10.1109/APEMC.2016.7522923
12. A. Routray, R.K. Singh, R. Mahanty, Capacitor voltage balancing in hybrid cascaded multilevel inverter using genetic algorithm at higher modulation indices, in *2018 IEEE Energy Conversion Congress and Exposition (ECCE)*, Portland, OR, 2018, pp. 3688–3693. https://doi.org/10.1109/ECCE.2018.8557986
13. A. Chatterjee, A. Rastogi, R. Rastogi, A. Saini, S.K. Sahoo, Selective harmonic elimination of cascaded H-bridge multilevel inverter using genetic algorithm, in *2017 Innovations in Power and Advanced Computing Technologies (i-PACT)*, Vellore, 2017, pp. 1–4. https://doi.org/10.1109/IPACT.2017.8245005
14. S. Joseph, C.A. Babu, Performance analysis of multilevel inverter with battery balanced discharge function and harmonic optimization with genetic algorithm, in *2016 International Conference on Next Generation Intelligent Systems (ICNGIS)*, Kottayam, 2016, pp. 1–6. https://doi.org/10.1109/ICNGIS.2016.7854068
15. K.M. Kotb, A.E. Hassan, E.M. Rashad, Implementation of genetic algorithm-based SHE for a cascaded half-bridge multilevel inverter fed from PV modules, in *2017 20th International Conference on Electrical Machines and Systems (ICEMS)*, Sydney, NSW, 2017, pp. 1–6. https://doi.org/10.1109/ICEMS.2017.8056071

Chapter 10
A Concept Design of a Futuristic Battery Management System for Submarines Using IEEE802.3bt Network

Arun Singh and Anita Khosla

Abstract This paper covers the concept design of futuristic battery management system (FBMS) for submarines using IEEE 802.3bt. Submarines all over the world use batteries for their operations including propulsion. The paper talks about various types of batteries used in submarines. These batteries pose a serious safety hazard if not managed properly. Therefore, a battery management system (BMS) is required to ensure safe operation of these batteries. The concept design will cover hardware and software design of a futuristic battery management system (FBMS). The FBMS will be used for management of batteries and controlling battery charger using IEEE 802.3bt network.

Keywords FBMS · BMS · Battery management system · Battery monitoring system · Power over ethernet (POE) for BMS

10.1 Introduction

Submarines are manned floating vehicles which can operate underwater. There are three main applications of submarines namely military, research, and tourism. Submarines for military purpose will be discussed in this paper. Submarines at sea serve on a combat footing. There are no national boundaries at sea. An armored battalion storming across a border can be fired upon by defenders, whereas a submarine contact at 1000 yards is just as dangerous, but must be given clear passage until it launches weapons [1].

We are entering a new warfare age, where stealth and information play key roles. In the center of this new military age lies the combat submarine. It played a dominant role in all the twentieth century's major conflicts, and its importance is still growing. At the end of the World War II, some 20-odd nations fielded submarines. Today, the number is 40-plus and growing [1].

A. Singh (✉) · A. Khosla
Faculty of Engineering and Technology, Manav Rachna International Institute of Research and Studies, Faridabad, India

A. Khosla and M. Aggarwal (eds.), *Smart Structures in Energy Infrastructure*, Studies in Infrastructure and Control,
https://doi.org/10.1007/978-981-16-4744-4_10

The submarine is penultimate force multiplier. One hundred men can comprise less than a company of infantry; or the same number can operate a nuclear submarine that will render a surface fleet impotent by its mere presence or demolish with a cruise missile strike the major command centers of a hostile nation. At the apex of power, one ballistic missile-armed submarine with 200 or less crew can destroy an entire civilization [1].

It takes craft and cleverness to build this force multiplier. Such craft and cleverness for concept design of one of the equipment of the submarine, i.e., battery management system, is the subject of this paper.

10.2 Role of Rechargeable Batteries in Submarines

All submarines use rechargeable batteries. Nuclear submarines must have them as an alternative source of power in case of total reactor failure to either restart the reactor or provide emergency limp-home propulsion. Diesel electric submarines rely upon them entirely for submerged operation and propulsion. For diesel electric submarines, the batteries occupy considerable space and, because of their weight, are installed in the lower part of the hull [2]. Improvements in batteries, including reduced weight and greater electrical power storage, will benefit diesel electric submarines to a great degree, and these advantages will prove useful to designers of nuclear boats as well [1]. Power requirement of battery for a submarine is generally in the range of MWh and can be calculated as given below:

- German type 209 batteries give 9400A-Hours per cell at a six-hour discharge rate [1].
- It uses 120 cells [1] of lead acid type which are connected in series. Thus, total voltage is 120 cells × 1.9 V (Each cell) = 228 V.
- Total energy = 9400 A-H × 228 V = 2.14 MWh.
- Submarines generally have two battery pits, and therefore, total energy will be 2.14 MWh × 2 = 4.28 MWh.

10.3 Various Types of Batteries Used in Submarines

Various types of rechargeable batteries which can be used in submarines are given (Fig. 10.1).

The **flooded lead acid** batteries are the primary source of energy in diesel electric submarines. As seen above, various variants of lead acid batteries, e.g., valve regulated [(Absorbent Glass Material (AGM)/Gel], flooded, etc., are used.

Lithium-ion batteries are also being used in the submarines. These batteries have advantage of low maintenance, devoid of discharging of gas, no filling of water,

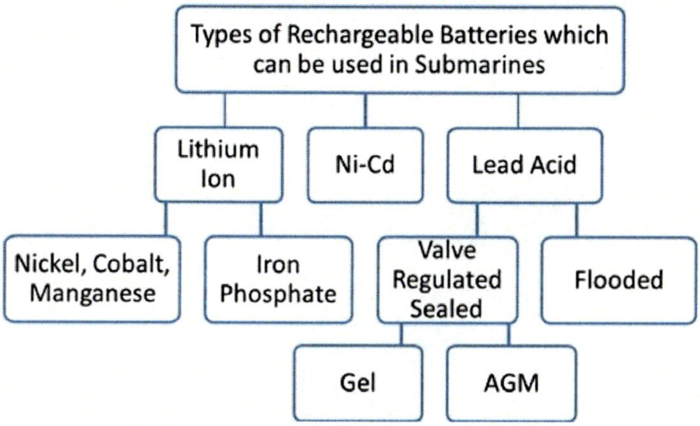

Fig. 10.1 Various type of rechargeable batteries for submarines

no manual measurements, no cleaning, etc. They provide higher submarine availability and need less replacements of battery because of their long lifetime. But these batteries need very close monitoring of the charge and discharge cycles, and therefore, a very reliable and rugged battery management system is required [3]. The battery with lithium-ion has high discharge current rate with higher energy efficiency than conventional batteries with NiMH/NiCd. However, threats to battery protection such as elevated fire hazards, flame, heat, and smoke are caused by these capabilities. By monitoring the battery parameters or the battery's functionality, the battery management system will reduce the risk from these vulnerabilities and provide safe operation of batteries onboard submarines. Typical battery-related safety and construction standards considered for the design of BMS do not resolve the possibility of electronic failure of the BMS for vehicles where these batteries are fitted. This inadequacy can be bridged by ISO 26262 process whose concepts can be used for protection of vehicles fitted with large battery packaging [4]. Variant of lithium-ion batteries are used in battery-powered vehicles across the world. The technology is evolving every day, and therefore, these batteries will become common storage devices for submarines in the future. Location of battery in a typical submarine is given [5] (Fig. 10.2).

10.4 Concept Design of a Futuristic Battery Management System

"**Battery management system** is needed to control the dynamics of energy storage process in the battery in order to improve the performance and extend battery life [6]". So far, various battery-powered devices use several forms of battery management systems which include battery-operated vehicles, big traction machineries in industries, mobile phones, ships, submarines, missiles, torpedoes, etc. Depending on the form, usage, and size of batteries, the architecture and design of BMS differs.

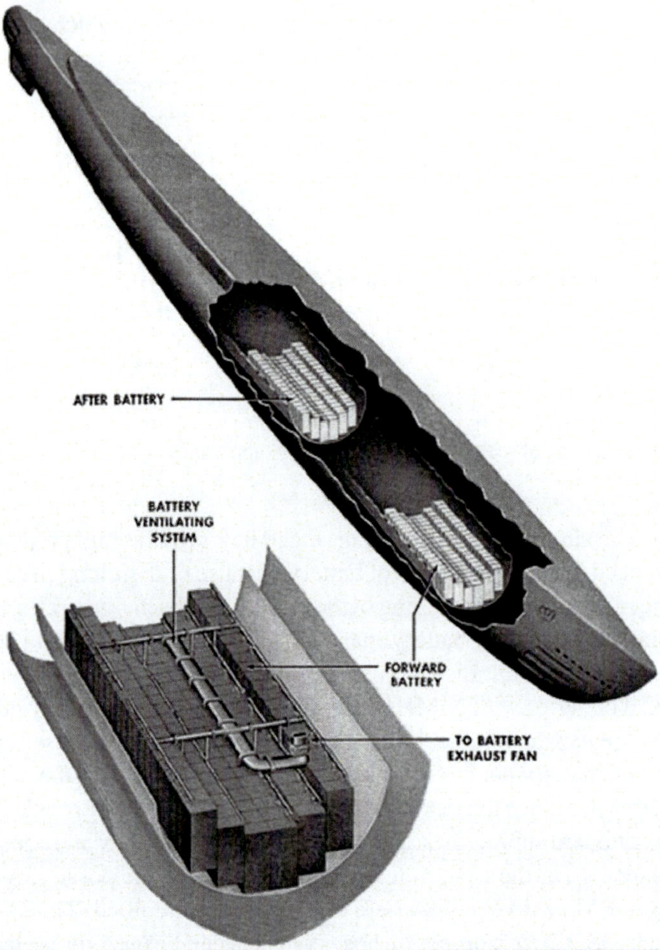

Fig. 10.2 Location of batteries in asubmarine

Challenges related to wiring in standard BMSs are being resolved by wireless battery management systems (WBMS). It is expected that the emerging cloud/edge computing technologies and Internet of Things (IoT) along with advances in WBMSs will result in providing significant value in cost reduction, extended scalability, and greater visibility in the lithium-ion battery energy storage systems. The WBMSs, however, are increasingly vulnerable to cyber-attacks due continuously link with networks. This can be overcome by block chain technologies to ensure connectivity and data protection against malicious cyber-threats by IoT-enabled WBMS [7]. On similar lines, electric bicycles also use the battery management system which tracks the battery pack's location data, energy storage, temperature, current, and voltage in real time, and updates information in cloud management program which can be used for optimizing utility of cycles by bicyclists [8].

Since BMS onboard submarine will be a mission-critical system and wireless technology has limited use onboard due steel hull and compartment separation through steel partitions, WBMS is not being considered. Battery management system (BMS) have also been designed for fast balancing throughout LiFePO4 (four lithium iron phosphate) cells charging using controller area networks (CAN) [9]. But this needs separate power supply connection for sensors which in the case of IEEE802.3bt is power over Ethernet. Therefore, futuristic battery management system will use IEEE 802.3bt whose functionality can be enumerated as given below:

- Battery data acquisition (temperature, current, voltage, electrolyte level) over IEEE802.3bt network.
- Provides state of health (SOH), battery residual life, state of charge (SOC), battery Ah, prediction of time duration a battery may discharge safely by collection of parameters, and their calculation as per set algorithms for a type of battery.
- Charge balancing
- Thermal and safety management communication data to a centralized control room/remote location through IEEE 802.3bt Ethernet including remote management of the system.
- Fault alarms and system status display.
- Protection against accidental mis-wiring and over voltage on all channel inputs.
- Sends battery parameters, alarm status, alerts during exigencies, battery summary through wired, and wireless network.
- Predicts and provides upcoming malfunction warnings based on temperature, current, and voltage set limits.
- Logging of cell/battery parameters (electrolyte level, current temperature, and voltage) and alarm conditions with time/date stamp.
- For data study and management, exports/Generates data.

Battery management system will take input from sensor of each cell. From these sensors data received will be analyzed and health of the battery will be estimated. Batteries will have different mode of charging. Following process is used for charging deep discharged batteries:

- **Constant Current Charge**. In this mode of operation, batteries are charged at constant current rate till a predefined voltage has reached. Battery voltage continues to increase until it exceeds the predefined level.
- **Constant Voltage Charge**. In this mode of operation, batteries are charged at constant voltage till batteries are fully charged. Charging current of the battery keeps reducing till minimum level of current is reached.
- **Trickle Charge**. In this way, a fully charged battery is charged at a rate equal to the self-discharge rate such that the battery stays at its maximum charged level.

Futuristic Battery Management System with Sensor in each cell of one set of 120 Cells forming Battery using IEEE802.3bt Network. Futuristic battery management system comprises hardware and software. Hardware is used for collecting data from sensors in each cell of battery and other interfaced devices

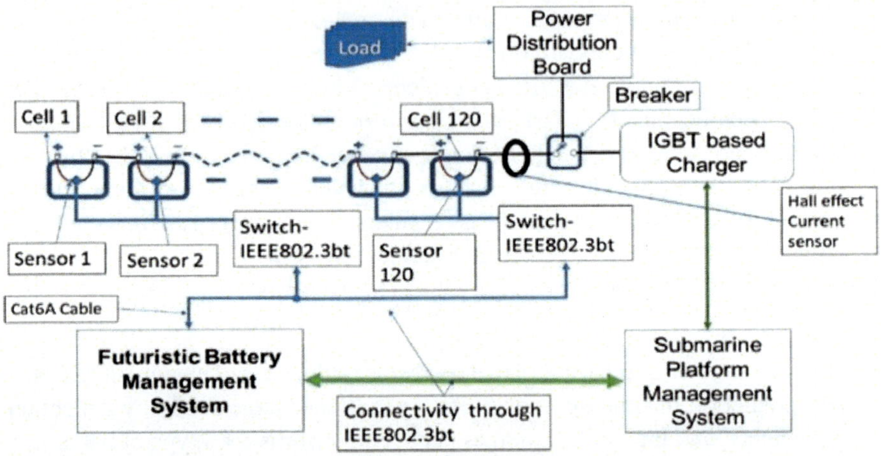

Fig. 10.3 Futuristic battery management system with sensor in each cell of one set of 120 cells forming battery using IEEE802.3bt network

for controlling the charging current for batteries as per algorithm executed by software to ensure safe operation of the whole system. FBMS connectivity and process of operation is shown in Fig. 10.3. FBMS is connected to submarine platform management system (SPMS) through IEEE802.3bt network. SPMS in submarines will be used for control and operations related to floating, moving, and fighting capabilities of all systems onboard. SPMS will also be used for controlling IGBT based charger for batteries. Batteries are either being charged from IGBT based charger controlled by FBMS and SPMS or connected to load through breaker and power distribution network for various loads onboard submarine.

Data from the sensor of each cell is acquired through a switch using IEEE 802.3bt network protocol. Depending upon the size and location of these cells, switches can be finalized. Since submarine batteries are large in size, and basic sensor will also be bigger in size compared to sensors for smaller batteries. For each set of 20 cells, a network switch can be provisioned, and thus, there will be six network switches for 120 cells. These network switches can be connected again through IEEE 802.3bt network to FBMS. Sensor will be fitted inside each cell and will need to be tailored to fit inside each type of technology of battery, e.g., Li-ion, lead acid, etc.

Sensor data can be transmitted by various protocols, e.g., hardwired, power line communication [10], MIL Bus 1553, etc. In this paper, this connectivity will be established through IEEE 802.3bt network which is a power over Ethernet and has advantage of no separate power supply requirement for cells or switches. Power source for IEEE 802.3bt network can be maintained in FBMS itself and fed through Cat 6A cables to individual cells and switches.

The challenge would be to integrate FBMS with universal sensor and SPMS for seamless operation in a submarine environment which can be overcome by following the development process for hardware and software which has been explained in

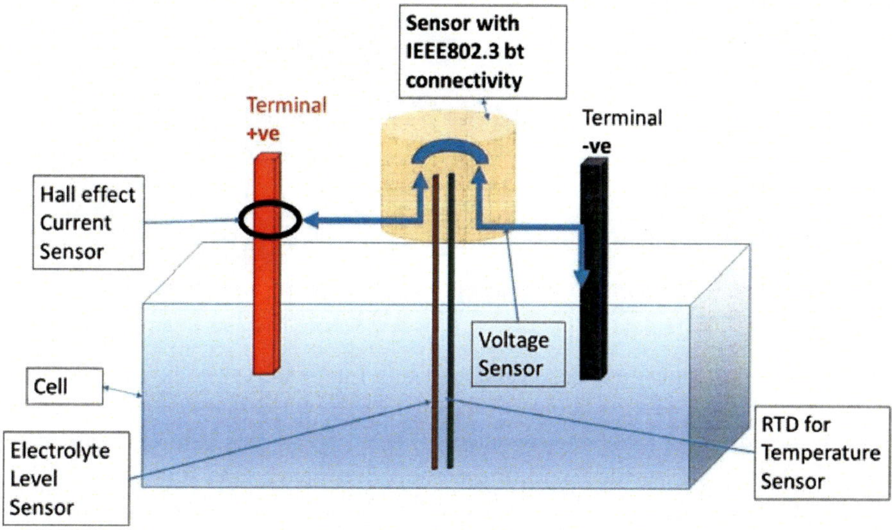

Fig. 10.4 Schematic of a sensor of a cell

succeeding sections. Proposed layout diagram of one set of cells of 120 comprising one battery pit is given below:

In the design of any battery management system, the most critical component is the sensor used in each of the cell. A simple schematic of this sensor is given in Fig. 10.4. Universal sensor which can measure temperature, voltage, current, and electrolyte level is to be used in each cell of the battery for assessment of charge status and battery health status. Each cell will have a mini-interface card which will convert temperature, voltage, current, and electrolyte level signals into IEEE802.3bt network protocol.

10.5 Hardware Design of a Futuristic Battery Management System

The compact VPX 3U based futuristic battery management device can be configured with an I/O card, processor card, and power supply card. The 3U VPX gives advantage of condensed size, power, and weight (SWaP) requirement for design of BMS. This also provides advantage of high computing and processing capabilities and high-speed interconnections for mission-critical applications. It is considered to give better performance than VME 6U. Following modules can be used for making FBMS:

The Curtiss wright made processor VPX3-1260 3U Intel Xeon Coffee Lake Single Board Computer (SBC) may be utilized which is rugged and features the 8th Gen Intel Xeon processor with integrated graphics for delivering all-in-one processing solution. The VPX3-1260 is designed to have more than 50 percent higher computing

Fig. 10.5 Pictorial view of SBC

capacity than previous four-core versions using Intel's first ever six-core processor. The VPX3-1260 is the first Intel SBC to deliver up to 40G Ethernet access providing quicker data transfer and better network efficiency than ever before. Its local SSD and NVMe storage provides 3 to 5 times productivity and a capacity of up to 16 times over conventional SATA SSD interfaces. SBC pictorial view is shown in Fig. 10.5 [11].

- **3U Open VPX Multifunction I/O Board -Model Open VPX 68G5 from NAI.** Up to three intelligent control modules can be configured to the new NAI Open VPX 3U Multifunction I/O and communications frame, the 68G5. This low-power/high-performance board, ideally suited for robust military, industrial, and commercial applications, provides off-shelf solutions that speed up the implementation of SWaP-C integrated systems. NAIS Configurable Open System Architecture (COSA) delivers the largest package density and flexibility on any I/O multifunction in the industry to pick over 70 smart I/O, networking, and Ethernet transfer features. Completely validated preexisting features can be combined in a variety of different ways. Each I/O has processing capacity, eliminating data processing requirement in Single Board Computer system (SBC) [12].
- Advantage of VPX-3U architecture with Curtiss wright made processor VPX3-1260 3U and Open VPX 68G5 from NAI used together is to provide very compact and high-speed FBMS which will be 50% smaller than the 6U VME architecture being presently used in other BMS.

10.6 Software Design of a Futuristic Battery Management System

Software development of FBMS is to be done using ISO/IEC/IEEE 12207 systems and software Engineering—software life cycle processes. It is an international standard for software life cycle processes.

ISO/IEC/IEEE 12207:2017 categorized the life cycle processes of software into four major process groups: technical processes, technical management, organizational project-enabling, and agreement. A number of subcategories are used in each of these four phase classes, for example; the key supply and acquisition activities (agreement); configuration (technical management); and disposal, maintenance, and operation (technical) are covered. Thus, design, development, implementation, testing, delivery, and post-delivery activities/processes for software aspects of futuristic battery management system should be done using IEEE 12,207 and associated standards including agile development model and data distribution services (DDS) as middleware to ensure reliable and rugged FBMS for submarine application.

10.7 Futuristic Battery Management System Interface with Sensors and Associated Devices Using Triple Redundant IEEE 802.3bt Network

Since FBMS will be a mission-critical system, triple redundant IEEE 802.3bt network interface with sensors and associated devices is proposed to ensure high degree of reliability. In IEEE 802.3bt, all pairs are used in the four-pair cable described in Fig. 10.6, allowing current to be distributed evenly. This enhances the total power capacity which can be transmitted over a single PoE cable, in addition to the higher data rate of up to 10GBASE-T. Cat 6A cable will be able to provide the required power and thermal performance with four balanced twisted pair cables. Pictorial view of Cat6A cable with four twisted pairs is shown in Fig. 10.7. Connectivity of Cat 6A cable using IEEE 802.3bt is shown in Fig. 10.6.

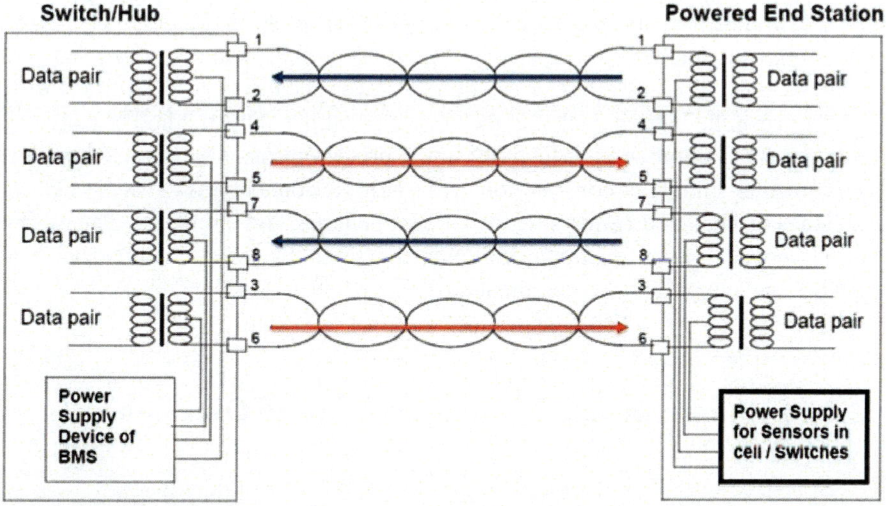

Fig. 10.6 Connectivity of CAT-6A cable using IEEE 802.3bt

Fig. 10.7 Pictorial view of Cat6A cable with four twisted pair

10.8 IGBT Based Charger Controlled via Submarine Platform Management System

As shown in Fig. 10.3, FBMS will get data related to all cells from sensors connected via IEEE802.3bt. Connectivity between sensor and network switches/FBMS will be as shown in Fig. 10.6. Algorithm for charging various type of batteries will depend upon their construction and capacity. Based on algorithm for type of batteries being used, BMS will decide the charging pattern and control IGBT based devices to charge the batteries via submarine platform management system.

For high-performance battery management systems for submarine applications, the IGBT based charger can be used which has voltage step up and cells balancing features within a single topology of converter. Several dual-active bridge (DAB) modules may be used in the device with supporting conversion principle. This conversion requires a smaller converter than that of traditional systems to be used as the converter only processes part of the output power according to the voltage difference between the battery pack and the converter output. This method of conversion often results in improved productivity in the processing of power at machine level. The converter's modular construction gives multiple connections to battery cells and facilitates cell balance during standstill, battery charge, and unloading. The machine will be regulated by a functional futuristic BMS that carries out cell balancing and manages the output voltage simultaneously [13].

10.9 Comparison and Advantages of FBMS Over Other BMS

As explained earlier, space is at premium in submarines, and therefore, equipment fitted inside it should be compact and small in size. Also, reliability of the equipment used onboard submarines has to be high because of their longer deployment below the sea. Keeping these in mind, the concept design of FBMS has following advantages for submarine application:

- IEEE 802.3bt can be used as universal interface among sensors, FBMS, SPMS, and other associated system, and therefore, multiple protocols, e.g., MILBUS 1553, RS 485, RS 232, etc., being used onboard can be avoided leading to reduction in size and complexity of equipment.
- IEEE 802.3bt will have a physical connectivity compared to wireless communication [7], and therefore, associated immunity against cyber-attack will be provided.
- IEEE 802.3bt with power over Ethernet connectivity will help in avoiding laying of separate cables for power supplies for sensors as done in the existing BMS, thus reducing number of cables required for system.
- Proposed universal sensor at Fig. 10.4 will avoid usage of multiple sensors in a cell leading to space saving and reliability by reducing number of components.
- Proposed hardware with VPX-3U connectivity will be very compact which will save space and provide high-speed connectivity compared to VME 6U being presently used in some of the BMS.

10.10 Conclusion

This paper has brought out concept of designing a futuristic battery management system for various type of rechargeable batteries onboard submarines which will keep changing due evolving technology and requirement to pack more energy in a constrained space. This concept design can be used as a universal battery management system for all types of rechargeable batteries in various submarines. Further action requires development of such a system, so that a futuristic battery management system of universal nature is available for management of rechargeable batteries for submarines across the globe.

References

1. S. Zimmermam, *Submarine Technology for the 21st Century*. ISBN 1-55212-330-8
2. R. Burcher, L. Rydill, *Concepts in Submarine Design*. ISBN 0 521 41681 7
3. https://www.udt-global.com/__media/libraries/sensors-and-processing/76---Anders-Wikstrom-Slides.pdf
4. S.S. Tikar, Compliance of ISO 26262 safety standard for lithium ion battery and its battery management system in hybrid electric vehicle, in *2017 IEEE Transportation Electrification Conference (ITEC-India)*, Pune (2017), pp. 1–5. https://doi.org/10.1109/ITEC-India.2017.8333870
5. https://maritime.org/doc/fleetsub/elect/chap5.htm
6. I.N. Haq, E. Leksono, M. Iqbal, F.N. Soelami, D. Kurniadi, B. Yuliarto, Development of battery management system for cell monitoring and protection, in *2014 IEEE International Conference on Electrical Engineering and Computer Science*, Bali, Indonesia (2014)

7. T. Faika, T. Kim, J. Ochoa, M. Khan, S. Park, C.S. Leung, A blockchain-based internet of things (IoT) network for security-enhanced wireless battery management systems, in *2019 IEEE Industry Applications Society Annual Meeting*, Baltimore, MD, USA (2019), pp. 1–6. https://doi.org/10.1109/IAS.2019.8912024
8. Y. Xu, S. Jiang, T.X. Zhang, Research and design of lithium battery management system for electric bicycle based on Internet of things technology, in *2019 Chinese Automation Congress (CAC)*, Hangzhou, China (2019), pp. 1121–1125. https://doi.org/10.1109/CAC48633.2019.8997319
9. L.A. Perişoară, I.C. Guran, D.C. Costache, A passive battery management system for fast balancing of four LiFePO4 cells, in *2018 IEEE 24th International Symposium for Design and Technology in Electronic Packaging (SIITME)*, Iasi (2018), pp. 390–393. https://doi.org/10.1109/SIITME.2018.8599258
10. M.S. Saleem, Development of PLC based communication architecture for battery management system, in *2020 IEEE 91st Vehicular Technology Conference (VTC2020-Spring)*, Antwerp, Belgium (2020), pp. 1–5. https://doi.org/10.1109/VTC2020-Spring48590.2020.9128451
11. https://www.militaryaerospace.com/directory/board-products/single-board-computers/product/14063159/curtisswright-defense-solutions-vpx31260-3u-vpx-intel-xeon-coffee-lake-sbc.
12. https://www.naii.com/3U-OpenVPX-Multifunction-I-O-Board-Model-68G5/P299
13. M. Shousha, A. Prodić, V. Marten, J. Milios, Design and implementation of assisting converter-based integrated battery management system for electromobility applications. IEEE J. Emerg. Select. Top. Power Electron. **6**(2), 825–842 (2018). https://doi.org/10.1109/JESTPE.2017.2736166

Chapter 11
Technological Advancements for Reduced Charging Time of Electric Vehicle Batteries: A Review

Abdullah Naim and Devendra Vashist

Abstract Electric vehicles require large energy storage for its operation. To meet this increased energy storage demand, batteries with different material combinations are developed. To develop next-generation batteries, improvements in the parameter that is cyclability, specific energy, charging rate, volumetric energy density, safety and stability are required. In the present work, different technological advancements in the field of electric vehicle batteries were studied. Factors affecting the charging speed of EV batteries are analysed in the research. Technology advancement in the area of battery chemistries, anode and cathode materials, charging methods that are taking place around the world, were summarized for improved performance of batteries.

Keywords Battery · Charging levels · Charging time · Lithium-ion battery · Charging stations

11.1 Introduction

An electric battery is used in full electric or hybrid vehicle to provide power to the electric motors used to run the vehicle. Commonly, these batteries are rechargeable and comprise of different combination sets such as lithium-ion, sodium-ion and NiCd. They are designed to provide high kilowatt-hour capacity. These batteries are different from lighting, ignition and starting batteries as it is designed to provide higher power output over a longer time interval and are of deep-cycle type. EV battery can be categorized on the basis of their weight-to-power ratio, energy and specific density. Compact size of the battery is more favourable as this helps in reducing unsprung weight of the vehicle thereby improving vehicle performance. In comparison with liquid fuels, most of them currently used batteries have lower specific gravity which affects the maximum range that can be achieved by the vehicle.

A. Naim · D. Vashist (✉)
Automobile Engineering Department, Manav Rachna International Institute of Research and Studies, Faridabad, Haryana 121010, India
e-mail: devendra.fet@mriu.edu.in

A. Khosla and M. Aggarwal (eds.), *Smart Structures in Energy Infrastructure*, Studies in Infrastructure and Control,
https://doi.org/10.1007/978-981-16-4744-4_11

The commonly used type of battery in the EV's are made of either lithium-ion or lithium polymer as they have a high energy-to-weight ratio. That means lighter batteries can be employed for covering a designed distance as they can store more energy per unit weight. EV batteries' other chemistries include lead acid, nickel–metal hydride, nickel–cadmium sodium chloride with nickel (zebra) batteries which are other less known chemistries available in the market [1].

11.1.1 *Types of Batteries Used in EVs*

11.1.1.1 Lead Acid Battery

Gaston Planté, a French physicist, invented the earliest type of rechargeable battery in 1859. Its cells are capable to supply high amount of current despite it having low energy-to-weight ratio. This quality along with its low cost makes it a suitable choice for being used in motor vehicles.

Since these batteries are inexpensive, they are widely used even in places where others can provide more energy. In 1999, their sales accounted for 40–45% of battery sold worldwide which was equivalent to $15 billion [2]. Hospitals, cell phone towers and renewable energy storage (solar/wind) use large size and number of these batteries to store backup power during power non-availability from grid. In their charged state, the charge is stored between PbO_2 on the positive terminal and pure Pb on the negative terminal, moreover an aqueous sulphuric acid as electrolyte. The production of energy while the battery is discharging is due to the release of energy when strong bond of $H^+$and O^{2-} ions are formed from water (H_2O) molecules of PbO_2.

11.1.1.2 Nickel Metal Hydride Battery

Invented in 1967, for providing energy needs in the form of battery for electric vehicles. It has a chemical reaction similar to that of NiCd battery for the positive terminal, with both the terminals having nickel hydroxide. The key difference being that the negative terminal uses alloy which absorbs hydrogen in place of cadmium. These batteries have three times the capacity as compared to NiCd battery, although less than lithium-ion batteries.

An intermetallic compound is used for metal (M) on the anode of the NiMH battery. Many developments have been made in this field, but the use of current falls in these classes. The most commonly used being AB_5, where A is a mixture of rare magnets like neodymium, cerium, lanthanum and praseodymium, and B is cobalt, nickel, aluminium or manganese. Certain cells use higher charge-holding ability materials which are based on AB_2 compounds, where A is either vanadium or titanium, and B is nickel or chromium with iron, manganese, chromium or cobalt [3].

The cells of NiMH batteries use potassium hydroxide as an alkaline electrolyte. The negative terminal has hydrogen in its interstitial metal hydride form, and the positive terminal has nickel hydroxide [4].

11.1.1.3 Lithium-Ion Battery

Akira Yoshino invented this chemistry in 1985, based on the pattern designed in the 1970s–1980s by John Good enough, M. Stanley Whittingham, Rachid Yazami and Koichi Mizushima [5–7]. It is a type of battery used in electric vehicles and portable electronics most commonly used for aerospace and military applications [8]. In these batteries, the charge flows from the cathode towards the anode through an electrolyte while discharging and back when charging. They use intercalated lithium as anode and graphite as cathode material. These batteries have low self-discharge, no memory effect [9] and high energy density. They are considered as a fire hazard as the electrolyte is flammable, which if damaged or improperly charged may lead to explosion. Based on safety-related issues, Samsung had to recall all of its Galaxy Note 7 handsets [10], and also there have been incidents involving the safety of battery of Boeing 787 aircraft. Table 11.1 shows a comparative chart of parameters for Li-ion and NiMH batteries.

Table 11.1 Comparison of Li-ion and NiMH battery

Battery name	Li-ion	NiMH	Units
Charge capacity (Max.)	75	85	Ah
Voltage	323	288	V
Energy	24.2	24.2	kWh
Voltage ratio (Max./Min.)	339/308	274/302	V
Charge	100	100	%
Number of cells per row	12	20	–
Number of rows	17	20	–
Internal resistance charge/discharge	1/1	1/1	Ω
Operating temperature	33	36	°C
Specific heat transition	0.4	0.4	W/K
Specific heat capacity	795	677	J/kg * K
Weight	318	534	kg

11.2 Battery Charging Methods

The electric vehicle charging station is a machine which supplies electric charge needed to recharge the batteries of fully electric and plug-in hybrid vehicles. Charging stations can be categorized into four main categories.

11.2.1 In-House Charging Stations

In these, the owner plugs in the vehicle in a normal wall outlet and use the car charger provided by the manufacturer to charge the vehicle battery overnight [11]. The home charging system requires a dedicated circuit wiring in order to be able to fast charge the vehicle [12].

11.2.2 Public Charging Stations

It is either a private or commercial undertaking which may be charged or is offered for free by the vehicle manufacturer. The chargers power output may be slow or high depending on the supply of power to the charger. Most of these public chargers are placed in the parking lots of shopping complexes which enable the EV owner to charge their cars which visiting the nearby facilities [13].

11.2.3 Fast Charging Stations

They have a power output of more than 40 kW and are capable of supplying over 100 km of range in 10–30 min of charging. They are usually placed at rest stops on highways to allow us to travel long distances. They are mainly used by commuters in metropolitan areas. They use Type2 and Type 3 connectors. Examples of such chargers include CHAdeMO and Tesla superchargers [14].

11.3 Types of Charging Levels

11.3.1 Alternating Current (AC) Supply

This type of charging structure is available both privately and publicly. It has the ability to connect directly an already available power structure of a home or business. It connects the in-built charging system of the EV directly to AC power supply.

AC current has two levels:

i. AC Level 1: It connects to the standard 120 V residential outlet directly and is able to supply 12–16 A (1.4–1.92 kW) which depends on the load-carrying capacity of the circuit. Almost every EV manufacturer includes this type of charger along with the vehicle
ii. AC Level 2: It uses 240 V (residential) or 208 V (commercial) power to supply an output of 6–80 A. It is used in the public charging stations as it provides better charging speed as compared to Level 1 charging with many more benefits.

11.3.2 Direct Current (DC) Supply

It is referred to as Level 3 charging. It connects the supply from grid to the EV battery by passing the grid power through an AC/DC Invertor.

DC current has two levels:

i. Level 1: It has a max. power supply of 80 kW at 50–1000 V.
ii. Level 2: It has a max. power supply of 400 kW at 50–1000 V.

11.3.3 High-Voltage Power Supply

It is the one which capable to give power outputs of hundreds and thousands of volt to charge the battery. It uses a special type of connector to provide a higher output while preventing arcing, breakdown of insulation, and accidental contact to humans. These connectors are generally used to provide current above 20 V, whereas other connectors are used for comparatively low voltages. Some of the high-voltage power supplies provide a digital as well as an analog communication system which can be used to manipulate and control the voltage output. This type of power supply is used to apply the bulk of input power into a power invertor which is connected to a voltage multiplier to produce a high voltage.

11.4 Charging Time

The charging time of the battery at a charging station depends on the charging rate provided by the charger and the capacity of battery being charged. The time taken by the charger to charge the battery depends on the type of charging level used to charge the battery. The charging level depends on the voltage that can be handled by the battery and the in-built charging system of the vehicle.

Charging time taken by the battery to charge can be calculated mathematically by using the formula.

Charging Time (h) = Battery Capacity (kWh)/Effective Charging Power (kW) [15].

The available battery capacity of the first generation of EV's like the original Nissan Leaf is 20 kWh providing 160 km of range. Tesla is the first ever company to mass produce long range EV's with battery capacities of 40kWh, 60kWh, and 85kWh having an estimated range of 480 km.

11.5 Factors Affecting the Charging Speed of EV Battery

The five main factors that affect the time taken by the EV battery to charge are:

i. Size of the battery: The size of battery is determined on the basis of its battery capacity. If the battery to be charged is of a larger capacity, then it will take a longer time to charge.
ii. State of the battery: If the battery which needs to be charged is completely empty, it will take longer time to charge as there is no current in the battery, and each cell will begin charging from zero.
iii. Charging rate of the battery: Each battery has a different charging rate as set by the manufacturer. If an EV is set to charge only till a rate of 7 kW, then it will not charge at a higher rate even at a fast car charger.
iv. Charging rate of the charging station: The time taken by the battery to charge will also be affected by the type of charger used to charge it and the maximum charging rate the said charger can provide. If the charging rate of a charger is 11 kW, but a car supports charging up to 20 kW, still the car will charge at the set rate of the charger.
v. Environmental factors: In colder temperatures, the battery tends to become cold, and this makes it difficult for the battery to absorb charge at a higher rate which increases the charging time of the battery and also decreases the efficiency of the battery.

11.6 Technological Advancements in the Field of EV Battery

The scientists are currently working on a number of technologies in order to make the EV battery more efficient and reduce their charging time. These developments include.

11.6.1 *Vertically Aligned Carbon Nanotube Electrodes*

NAWA technologies [16] have developed a new pattern for the electrode of EV batteries and are calling it the ultra-fast carbon electrode, which uses vertically aligned carbon nanotubes which are said to increase the power output from battery by ten times, and also increasing its energy storage by three times to the ones being used currently and also increase the life cycle of the battery to five times. The company aims to achieve 80% charge in just 5 min and a range of 1000 km.

11.6.2 *Lithium-Ion Battery (with no Cobalt) [17]*

University of Texas researchers have developed a lithium-ion battery which uses nickel instead of cobalt for its cathode. This has made the battery easy and inexpensive to manufacture. Moreover, this battery claims a better battery life and even distribution of ions.

11.6.3 *Ryden Dual-Carbon Power Storage Unit/Battery [18]*

Power Japan Plus is in the process of developing a new battery technology called Ryden Dual Carbon, which they claim increased battery life and gives faster charging. They claim that with this new technology, we would be able to charge our batteries twenty times faster than lithium-ion batteries, and it will also increase its durable and last for more than 3000 charging cycles. They also claim these batteries to be safe with low chances of fire and explosion.

11.6.4 *Silicon Anode Lithium-Ion Batteries [19]*

University of Eastern Finland researchers have found a solution to the issue of unstable silicon in Li-ion batteries by producing a hybrid negative terminal made of carbon nanotubes and mesoporous silicon microparticles. The main reason for replacing graphite with silicon is its higher charge holding capacity and that it has shown to increase the battery capacity by ten times.

11.6.5 Lithium–Sulphur Battery [20]

Monash University scientists have developed a lithium–sulphur battery; they claim to have a lower environmental impact than lithium-ion. The developed system require less maintenance cost, while offering a vehicle range of 900–1000 km.

11.6.6 IBM Salt Water Batteries [21]

IBM has identified a new battery chemistry that discards heavy metals and has higher performance as compared to lithium-ion battery. The raw material required for this system can be extracted from sea water thereby making this battery cheaper to manufacture. Another important feature is that it can be charged faster than lithium-ion. They claim that the technology is appropriate for EV's and have already started to work on this project with Mercedes -Benz to develop a viable product.

11.6.7 Panasonic Battery Management System [20]

Panasonic has developed a new technology which can easily help us monitor and govern the value of lithium-ion particles in the battery very easily. This system can be applied to any battery which will help us to drive towards a sustainable future and also allows better management, reusability and recycling of lithium-ion batteries.

11.6.8 Silanano Sand Batteries [22]

These batteries use silicon to get thrice the amount of performance which we get from the currently used graphite Li-ion batteries. But silicon is degrading drastically and is expensive to be produced on a large scale. But sand can be used to purify, powder mixed with magnesium, and salt before heat is supplied, so that oxygen can be removed. This process results in the formation of pure silicon. This has helped increase the performance and life-span of batteries. Silanano is a start-up working on battery chemistries and about to commercialize this technology for EV market. The company is currently working alongside companies like Daimler and BMW.

11.6.9 Solid-State Lithium-Ion Batteries [23]

Toyota has developed a solid-state battery using sulphide superionic conductors, which allows the battery to work as super capacitor. This techniques helps it to completely charge in 7 min making storage system ideal for cars. It is said to be able to work in extreme temperature conditions of as low as negative 30 °C and up to 100 °C. It is still a work in progress and will take a little time to reach the current market.

11.6.10 Sodium-Ion Battery [20]

Scientists in Japan are developing new battery technology that does not require lithium but use sodium which is seven times more efficient than conventional batteries. These batteries are said to arrive in the next five to ten years.

11.6.11 Carbon-Ion Battery [20]

UK based organization ZapGo has already technologically advanced and manufactured the first-ever carbon-ion battery. It has many benefits like superfast charging and much better performance than lithium-ion batteries with still having the ability to be fully recyclable.

11.6.12 Safer and Faster Charging of Current Lithium-Ion Batteries [24]

Scientists at University of Warwick have designed a new charging system for the lithium-ion battery. The system is developed in such a faction that battery charges 5 times faster than the current approved limits. The technology constantly measures the temperature of the charging battery and provides the charge relative to it without allowing the battery to ever overheat.

11.6.13 NIO Power Swap [25]

NIO is a Chinese electric vehicle manufacturer which along with its vehicles has developed a solution for reduced charging time by removing the hassle of plugging in your battery out of the system. They call this technology NIO Power Swap in which the battery of the vehicle gets swapped in less than 3 min all the while keeping the system in check and in good shape that too at no extra cost.

11.6.14 Ion Energy [26]

It is a Mumbai-based Indian start-up company which takes a software approach in order to improve battery life and performance. The company has developed a battery intelligence platform and that had named it Edison Analytics. This will help in identification of the reasons for degradation and suggests corrective measures to increase battery life.

11.6.15 Lohum Cleantech [26]

It is a Delhi, India, based company which helps in full recycling of used Li-ion batteries. They extract materials such as cobalt, graphite, manganese sulphate and nickel sulphate, which can further be used to produce new Li-ion cells.

11.6.16 Gegadyne Energy [26]

The company started in 2015 and has patented their product which offers longer life cycle and high energy density than their counter part, i.e., Li-ion battery. The company also claims 0–100% recharging in 15 min.

11.6.17 Grabat Graphene Batteries [20]

Grabat has manufactured grapheme batteries which offer a driving range of 500 miles in one single charge. They also claim a 33 times faster charging rate than Li-ion batteries.

11.6.18 Sila Nanotechnologies [27]

Scientists at Tel Aviv University have developed a low-cost silicon nanostructured anode. Beacuse of this, safety parameters have improved. This technique has also resulted in achieving 20% more energy density over the current existing battery chemistries.

Table 11.2 provides a comparative details of the different ecologies that will play a major role in future battery technologies.

11.7 Conclusion

Configuration of current and future EV battery technologies was analysed, wherein it concluded that current technologies need improvement, lot of scope is available, and market forces are also supporting this new trend. New developments in batteries in this area of carbon nanotubes, dual carbon, IBM's salt water technology, and solid-state lithium-ion seems to be more viable options which can fulfil the current energy storage needs of the EVs.

Table 11.2 Comparison of developing technologies for EV batteries

S. no.	Technology type	Developer	Feature 1	Feature 2	Feature 3
1	Vertically aligned carbon nanotube electrodes	NAWA technologies	Increase in its energy storage by three times	Increase the life cycle of the battery to five times	80% charge in 5 min with 1000 km range
2	Cobalt-free lithium-ion battery	University of Texas	Easy and inexpensive to manufacture	Increased battery life	Even distribution of ions
3	Ryden dual-carbon battery	Power Japan plus	20 times faster charging than lithium-ion batteries	3000 + charge cycles	Safer
4	Silicon anode lithium-ion batteries	University of Eastern Finland	Higher charge holding capacity	Increased battery capacity by ten times	Reduced cost
5	Lithium–sulphur battery	Monash University	Less maintenance cost		
6	IBM salt water batteries	IBM	Extracted from sea water thus cheaper to manufacture	Fast charging	
7	Battery management system	Panasonic	Monitoring and determining the value of lithium-ion in the battery very easily	Increased safety	
8	Sand batteries	Silanano	Increased performance	More life-span	
9	Solid-state lithium-ion batteries	Toyota	Completely charge in 7 min	Temp. range − 30 to 100 °C	
10	Sodium-ion battery	Japan	Cheaper	Easy availability of raw material	
11	Carbon-ion battery	ZapGo	Superfast charging	Better performance than lithium-ion	Completely recyclable

(continued)

Table 11.2 (continued)

S. no.	Technology type	Developer	Feature 1	Feature 2	Feature 3
12	Safer and faster charging of current lithium-ion batteries	University of Warwick	Charge 5 times faster	Technology constantly measures the temperature of the charging battery and provides the charge relative to it without allowing the battery to ever overheat	Safety
13	NIO power swap	China	Battery swap	No wastage of time	Free of charge
14	Ion energy	India	Improved battery performance and life	Measures to prevent battery degradaion	
15	Lohum Cleantech	India	Recycling of Li-ion batteries		
16	Gegadyne energy	India	Higher energy density	Longer life	Fast charging
17	Grabat graphene batteries	Grabat	Range of 500 miles in single charge	33 times faster charging	
18	Sila nanotechnologies	Tel Aviv University	Higher battery capacity	Safer structure	Higher energy density

References

1. Axeon Receives Order for 50 Zebra Packs for Modec Electric Vehicle; Li-Ion Under Testing. Green Car Congress (2016). Retrieved 15 Dec 2019
2. D. Linden, T.B. Reddy (eds.), *Handbook of Batteries,* 3rd edn. (McGraw-Hill, New York, 2002), p. 23.5. ISBN 978-0-07-135978-8
3. J. Kopera, *Inside the Nickel Metal Hydride Battery.* Cobasys. Archived (2009) Retrieved 2011–09–10
4. Nickel Metal Hydride Handbook (PDF) (NiMH02.01 ed.). Energizer Battery Manufacturing (2019)
5. IEEE Medal for Environmental and Safety Technologies Recipients. IEEE Medal for Environmental and Safety Technologies (Institute of Electrical and Electronics Engineers, 2019). Retrieved 29 July 2019
6. Jump up to: a b The Nobel Prize in Chemistry 2019 . Nobel Prize. Nobel Foundation (2019). Retrieved 1 January 2020
7. Jump up to: a b NIMS Award Goes to Koichi Mizushima and Akira Yoshino (National Institute for Materials Science, 2016). Retrieved 9 April 2020

8. M.S. Ballon, Massie Santos, *Electrovaya, Tata Motors to Make Electric Indica*". cleantech.com (2010). Archived from the original on 9 May 2011. Retrieved 11 June
9. Memory effect now also found in lithium-ion batteries. Retrieved 5 August (2015).
10. Jump up to:a b Kwon, Jethro Mullen and K. J., Samsung is recalling the Galaxy Note 7 worldwide over battery problem. CNNMoney (2016). Retrieved 13 September (2019)
11. Charging at Home. Energy.gov. Retrieved 3 October (2019)
12. P. Stenquist, Electric chargers for the home garage. The New York Times (2019). Retrieved 3 October (2019)
13. J. Savard, Is it time to add electric vehicle charging stations to your retail shopping center? Metro Commercial (2018). Retrieved 3 October (2019)
14. A Simple Guide to DC Fast Charging. Fleetcarma.com. Archived from the originalon 2017–12–26. Retrieved 2017–10–05 (2017)
15. Guide to buy the right EV home charging station. US: Home Charging Stations. 2018–01–03. Retrieved 2018–09–01 (2018)
16. Technology—NAWA Technologies. Retrieved 2021–12–02
17. The University of Texas at Austin; New Cobalt-Free Lithium-Ion Battery Reduces Costs Without Sacrificing Performance; 2020–14–07
18. E. Luke, Ryzen dual carbon battery charges twenty times faster than lithium ion, lasts longer, due this year. 22–05–2014 Ryden dual carbon battery charges twenty times faster than lith (pocket-lint.com). Retrieved 2021–12–02
19. T. Ikonen, N. Kalidas, K. Lahtinen, et al., Conjugation with carbon nanotubes improves the performance of mesoporous silicon as Li-ion battery anode. Sci. Rep. **10**, 5589 (2020)
20. H. Chris, Future Batteries, Coming Soon: Charge in Seconds, last months a (pocket-lint.com); (2020–10–07); Retrieved 2021–12–02
21. Y.-H. Na, Free of heavy metals, new battery design could alleviate environmental concerns. IBM Research Blogs, 2019–12–18, Free of Heavy Metals, New Battery Design Could Alleviate Environmental Concerns | IBM Research Blog Retrieved 2021–12–02
22. E. Luke, Get three times more battery life by using sand, 2014, Get three times more battery life by using sand—Pocket-lint (pocket-lint.com)
23. Y. Kato, S. Hori, T. Saito, K. Suzuki, M. Hirayama, A. Mitsui, M. Yonemura, H. Iba, R. Kanno, High-power all-solid-state batteries using sulfidesuperionic conductors. Nat. Energy (2016)
24. A. Tazdin, Understanding the limits of rapid charging using instrumented commercial 18650 high-energy Li-ion cells. Electrochim. Acta (2018)
25. NIO Power (2021)
26. Most Innovative Indian Start-ups working on Battery Technology. EV reporter (2021)
27. Electric Car Battery Technology Breakthrough—Latest Developments in Battery Technology—Vehicle suggest (2021)

Chapter 12
Compact and Efficient Way of LSEV Charging

Rakesh Sharma and Anita Khosla

Abstract This paper discusses about the value preposition, form factor and efficiency improvement of the low-speed electric vehicle (LSEV) chargers. The goal is to understand what are the different charging under standards, charger topologies and which topology is best suited for LSEV charger. Seeing the current market demand for smaller form factor, efficient, and cost-effective chargers, one has to choose a topology that meets the end user demands.

Keywords Off-Board LSEV charger · PFC · DC-DC converter topologies · EV · EVSE · Efficiency · Resonant converter

12.1 Introduction

The problem of air pollution is increasing day by day and has led to the need for development of clean and fuel-efficient transportation. Seeing this requirement, electric vehicles (EVs) appear to be the best choice over conventional IC engine-based vehicles. A light EV (LSEV) charger or low-speed EV (LSEV) charger or electric vehicle supply equipment (EVSE) or charging equipment are prerequisite for EV application by LSEV owners. LSEVs (two/three wheeler) are generally used for urban mobility with top speed limit of 50–60 km/h for two wheelers and 25–35 km/h for three wheelers.

As it is known that future of automobile industry and the way how people commute is changing with evolution of electric vehicles whether it is an industrial vehicle or commercial vehicle. To cover the masses and change their way of travel and last mile connectivity, electric two wheelers and electric three wheelers (usually known as E-rickshaw) are playing a vital role, and these two types of LSEV have seen

R. Sharma (✉) · A. Khosla
Manav Rachna International Institute of Research and Studies, Faridabad, Haryana, India
e-mail: rakesh_sharma2019@manavrachna.net

A. Khosla
e-mail: anitakhosla.fet@mriu.edu.in

A. Khosla and M. Aggarwal (eds.), *Smart Structures in Energy Infrastructure*, Studies in Infrastructure and Control,
https://doi.org/10.1007/978-981-16-4744-4_12

the highest surge in R&D, production, and evolution of new players in the market. LSEVs are emerging out to be the first choice for last mile connectivity, and electric two wheelers are also on the rise as new players are stepping in.

All LSEVs/EVs have electric motor as the prime mover and these motor are powered by the energy stored in the rechargeable batteries, and these batteries require DC power, which in turn can be given through LSEV/EV chargers. LSEV chargers can be broadly classified as AC chargers and DC chargers. DC chargers are usually required for fast charging of LSEVs; they convert the power before it enters the LSEV. The converter power directly goes into the car battery, bypassing the vehicle's converter system. DC chargers are not covered in the scope of this article.

Residential/Consumer level charging is convenient and inexpensive. Usually, people like to charge LSEV at home [1] over night using off-board chargers. Moreover, charging at home is cost effective as compared to public charging. AC chargers can convert regular home/utility AC supply to DC, and this DC can be fed to batteries. Utility supply at homes in India is 230 V_{ac}/15 A (3.5 kW AC output), and same can be used along with AC–DC charger to charge a LSEV like a two wheeler or a three wheeler with a power requirement of < 3.5 kW. For example, if a two wheeler have 3.6 kWh battery and AC–DC charger with an output of 1.2 kW, then it will take 3 h to charge the two Wheeler.

LSEV chargers discussed here is a Level 1 Type 2 EVSE charger complying to Bharat EV charging specifications for normal AC charging.

12.2 Types of Chargers and Charging Standards

As discussed above, there are broadly two type of EV chargers namely [2], AC chargers and DC chargers, and these can be further categorized into rapid chargers (40–350 kW), fast chargers (7.0–22 kW), and slow chargers (150 W–6.0 kW). Rapid and fast chargers are fed from three-phase AC supply, and on the other hand, slow chargers can be used with single-phase AC supply. There are different standards [3] available for EVs charger and connectors like IEC 61851, IED 60309, Bharat EV AC001, and for connectors like CHAdeMO, CCS, Type 2, etc. Use of a particular standard in a country basically depends on the discretion of local governing authorities. Some countries have not laid out the road map for standardization, and some countries are already working on the mandates for implementing (LS) EVs and related standards.

12.3 Pain Points for LSEV Users

(a) LSEV users usually do not have enough access to efficient charging.
(b) Lower driving range due to on-board battery capacity.

(c) Higher owning price of LSEV as compared to similar ICE (Internal Combustion Engine) vehicles.
(d) Lower charging speeds.

12.4 Key Benefits of Using Off-Board LSEV Chargers

(a) Weight on the LSEV is removed.
(b) Faster charging as on-board LSEV chargers are usually low power rated versions to optimize cost, weight, and space.
(c) Charging at higher power level with higher power chargers.

12.5 Topology

Depending on the type of charging power requirements (namely rapid, fast, and slow), chargeable time, and battery capacity, there can be different type of topologies that can be employed to design a charger. As the goal in this article is to design a LSEV charger capable of charging an electric two wheeler or electric three wheeler, so the discussion will revolve around this only. Selection of a topology for a particular charging application depends on many factors, and some key factors are presented here under.

12.5.1 Power Density

It is defined by how much power per cubic inch can be obtained by keeping the charger performance specifications intact. It is basically the system power to system volume in cubic inch. This defines the reliability and conciseness of the design. One has to balance between power density and required efficiency and thermal performance.

12.5.2 Efficiency

It is the measure of output power over the input power or it can be put like this, and it is the efficient conversion of input power to output power. The more is efficiency of charger, the less energy units are consumed for same output power to charge a LSEV. Efficiency is impacted by the losses incurred because of power loss in resistive components, switches, diodes, inductors, and transformers, these are the major contributors of losses in a design. So optimizing the values of these components and using right grade/technology can help to improve the system efficiency.

12.5.3 *Output Voltage and Current*

The output characteristics should be known for selecting a topology. It is the output power requirement and peak power demand that defines the topology selection, if capable of delivering the required power. For example, a flyback converter cannot deliver more than 150 W efficiently else becomes costly design to be productionized and a LLC resonant converter can deliver much higher power.

12.5.4 *Power Factor and Input Current THD*

These are required for efficient performance of electrical system and grid. If PF and iTHD are good, then the design is less susceptible to harmonics, and the utility line will also not suffer from stress due to low PF. Henceforth, integrating a PFC stage ahead of DC–DC stage is wise idea. By having a dedicated PFC stage, one can ensure tight regulation of output parameters and improved performance of the DC–DC stage.

12.6 LSEV Charger Design

The objective is to design a compact and efficient charger for normal AC charging keeping electric two wheeler and three wheeler in mind. So for this application, usually the charger power ranges anywhere between 300 and 3500 W (depending on the electrical vehicle type and motor/battery capacity). As per per current market trend, a charger with output of 48 V and 20 A is sufficient enough to charge both electric two wheeler and three wheeler. The topology that fits well in terms of form factor, efficiency, and power density (keeping price effectiveness in mind) is CCM boost [4] as first stage for power factor correction.

12.6.1 *The Different Type of PFC Boost Topologies*

12.6.1.1 CCM Boost PFC Topology

The diode bridge made from $D1$ to $D4$ is used to rectify the AC main input voltage to DC. Components L_{BOOST}, M_{BOOST}, D_{BOOST}, C_{o} form the CCM boost section [5]. In this topology, the difference between the diode current and the DC output current is high, and the ripple in output capacitor is also high. For higher powers, the diode bridge rectifier is not a viable option to use as losses are increased for high power operation. This topology can be used typically up to 1.5 kW of power, or in other

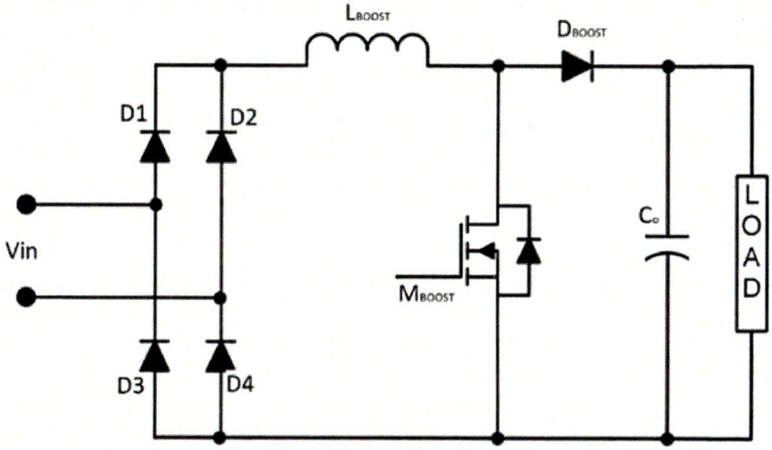

Fig. 12.1 CCM PFC boost topology

words, it can be said that this topology is best suited to low power to mid power range (Fig. 12.1).

12.6.1.2 Semi-bridgeless PFC Boost Topology

Here, two boost stages are required to achieve PFC [6, 7]. The boost inductors are connected to the converter's input. It requires full wave rectifier (*D*1, *D*2, *D*3, and *D*4) to charge the common PFC boost capacitance (C_o) to peak value during initial startup. Once the converter is operational, then it needs only one diode conducting (*D*1 or *D*2) instead of conventional two diode conduction. It is not easy to design though it offers high efficiency at low line and light load conditions. Using this topology offers benefits of improved charging time, form factor, reduced component count, and reduced EMI footprint (Fig. 12.2).

12.6.1.3 Bridgeless PFC Boost Topology

In this type of topology, the diode bridge rectifier is not required, and it gives the performance of conventional boost type topology. During the positive half cycle, the power is processed by the combination of *M*1 and *D*1. The current returns through *M*2 which act as rectifier. During the negative half cycle, the boost action takes place *M*2 and *D*2, and *M*1 gives the return path to current. Compared to the conventional boost PFC, one diode is eliminated from the circuit operation, resulting in less conduction losses. It is a good solution for mid–high power range where efficiency and power density are key design criteria. Since it resolves the thermal issue of diode-based bridge rectifier, but it introduces EMI into design. Since the input is floating with respect to PFC ground, it becomes almost impossible to sense the input mains. One

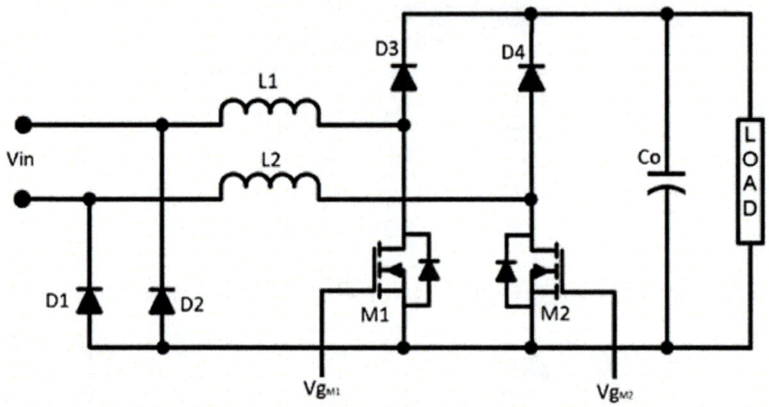

Fig. 12.2 Semi-bridgeless PFC boost topology

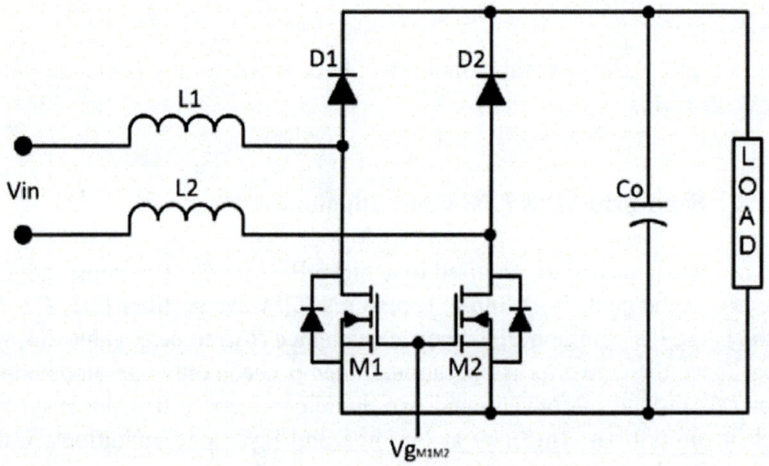

Fig. 12.3 Bridgeless PFC boost topology

has to make use of low frequency transformer or opto-coupler to do so. Both $M1$ and $M2$ can be driven by the same gate drive circuit, which simplifies the design (Fig. 12.3).

12.6.1.4 Interleaved PFC Boost Topology

In this topology [7, 8], two boost stages are interleaved. In this method, two PFC boost stages operate 180° out of phase. The sum of two boost inductor currents results in input current. This topology has a major benefit of input and output inductor ripple current cancellation because inductor ripple current are 180° out of phase. With the help of this topology, the size of boost inductor can also be reduced and EMI signature

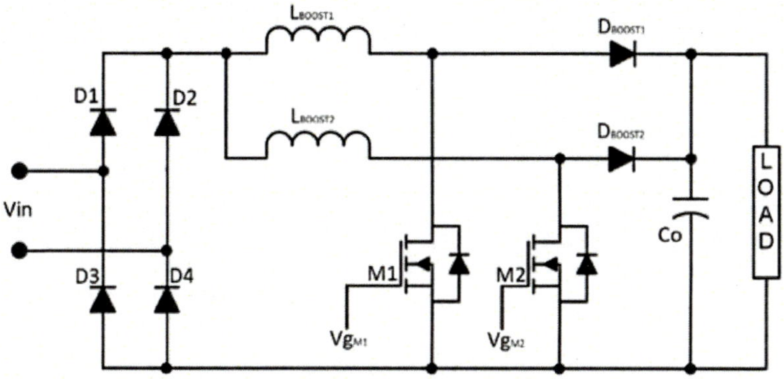

Fig. 12.4 Interleaved PFC boost topology

can also be improved. The boost capacitor RMS current is also reduced because of output inductor current ripple cancellation, and this helps in reducing the size of boost capacitor as well. Conduction losses can also be reduced to 50% as compared to conventional boost topology. The interleaved PFC's efficiency during light load condition can be improved by turn-off of a phase under light load condition, resulting in a light load efficiency similar to that of conventional pre-regulator (Fig. 12.4).

12.6.2 The DC–DC Secondary Stage Topologies

There are different topologies available for DC–DC high power stages [9], namely

(a) Half bridge converter
(b) Full bridge converter
(c) ZVS phase shift full bridge converter [10, 11]
(d) LLC resonant converter.

One of the above-mentioned topologies can be employed for the target application requirement. In view of the system performance (higher power density and high efficiency), the favorable choice is LLC resonant converter and same has been discussed here forward.

12.6.2.1 LLC Resonant Converter

LLC [12, 13] refers to the use of two inductances (L) and one capacitor (C). The half bridge switches $M1$ and $M2$ are turned on and off alternatively by 50% duty cycle square waves. This leads to an input voltage, which is an AC square wave that excites the circuit. L_s and C_s are connected in series with primary winding of

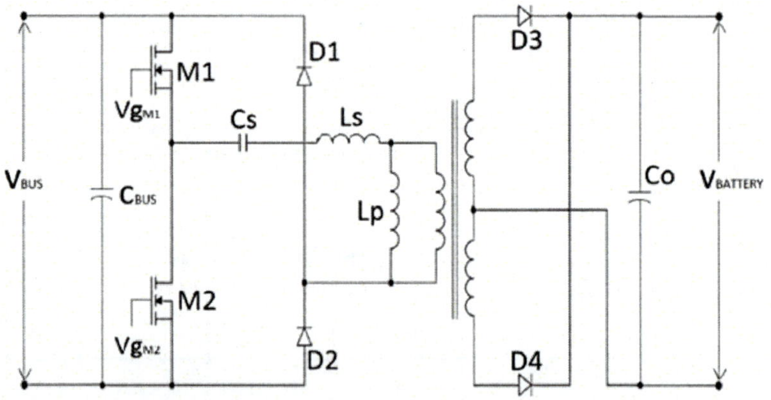

Fig. 12.5 LLC resonant topology

transformer (see Fig. 12.5). The magnetizing inductance L_p appears in series with the L_s, increasing the overall inductance.

The L and C values are chosen such that the resonance happens at the switching frequency. While the applied voltage is an AC square wave, the primary circuit is resonant and "rings" or oscillates at the resonant frequency, and the current is sinusoidal. This combination of L, L, C establishes a resonant at the switching frequency. As a result, the switching devices see a sine wave and are enabled to switch at zero-crossing points or near zero points. In consequence to this, switching losses are reduced. The benefit of doing so it that it allows for higher switching frequencies, which in turn, helps in reducing the size of filters and transformer as well as minimizing the switch heat dissipation and heat sinks. All of these are achieved with the increase in efficiency too.

By operating PFC and DC–DC stage at higher switching frequency, the size of inductors and transformers can be reduced and thereby achieving good form factor and power density, but it is not so easy to achieve. When the switching frequency of switching device's is increased then issues related to ground shifts, oscillations because of stray inductance and device output capacitances, and higher unwanted EMIs can crop up. By employing a good design practice and right selection of device and its technology, one can optimize the designs, and the mentioned issues can be rectified to great extents. A good example of device technology is Cool SiC device from Infineon technologies, this is MOSFET based on silicon carbide technology, and it is a great performance by improving efficiency and maintaining performance at high switching frequency operation. Similarly, with use of SiC Schottky diodes and better inductor and transformer designs, one can improve the charger efficiency and also by adding secondary side synchronous rectification can also help to improve the design performance.

12.7 Conclusion

To design a compact and efficient off-board LSEV charger [14, 15], one can employ a CCM boost or semi-bridgeless topology for PFC, and for DC–DC converter secondary stage, one has to employ a LLC resonant controller with secondary side diode synchronous control. Form factor can be improved by increasing switching frequency to operable limits and hence reducing the size of key components like inductors, transformers, and using SMD (surface mounted devices) semiconductor devices. A good control over output parameters and maintaining the required efficiency can be achieved the following the above key guidelines for design of compact and efficient LSEV charger.

References

1. https://www.zap-map.com/charge-points/
2. Understanding the different EV charging levels. https://evcharging.enelx.com/eu/about/news/blog/552-ev-charging-connector-types
3. Electric Vehicle Charging Standards and Infrastructure. https://www.allaboutcircuits.com/technical-articles/electric-vehicle-charging-standards-and-infrastructure/
4. B. Singh, B.N. Singh, A. Chandra, K. Al-Haddad, A. Pandey, D.P. Kothari, A review of single-phase improved power quality AC-DC converters. IEEE Trans. Ind. Electron. **50**, 962–981 (2003)
5. S. Abdel-Rahman, F. Stückler, K. Siu, *PFC Boost Converter Design Guide*
6. Proposed PFC Semi Bridgeless phase shifted boost converter fed four quadrant DC drive. https://shodhganga.inflibnet.ac.in/bitstream/10603/141545/12/12_chapter%204.pdf
7. M. O'Loughlin, *Choosing Between Semi-Bridgeless and Interleaved PFC Pre-Regulators*. https://www.powerelectronics.com/technologies/power-management/article/21861341/choosing-between-semibridgeless-and-interleaved-pfc-preregulators
8. Y. Jang, M.M. Jovanovic, Interleaved boost converter with intrinsic voltage-doubler characteristic for universal-line PFC front end. IEEE Trans. Power Electron. **22**, 1394–1401 (2007)
9. Y.J. Lee, A. Khaligh, A. Emadi, Advanced integrated bidirectional AC-DC and DC-DC converter for plug-in hybrid electric vehicles. IEEE Trans. Veh. Technol. **58**, 3970–3980 (2009)
10. R. Redl, L. Balogh, D.W. Edwards, Optimum ZVS full-bridge DC/DC converter with PWM phase-shift control: analysis, design considerations, and experimental results, in *Proceedings of 1994 Applied Power Electronics Conference and Exposition*, vol. 1, pp. 159–165
11. Y. Jang, M.M. Jovanovic, Y.-M. Chang, A new ZVS-PWM full-bridge converter. IEEE Trans. Power Electron. **18**, 1122–1129 (2003)
12. R. Beiranvand, B. Rashidian, M.R. Zolghadri, S.M.H. Alavi, Using LLC resonant converter for designing wide-range voltage source. IEEE Trans. Ind. Electron. **58**, 1746–1756 (2011)
13. W.-Y. Choi, B.-H. Kwon, J.-S. Lai, A hybrid switching scheme for LLC series-resonant half-bridge DC-DC converter in a wide load range, in *Proceedings of 2010 IEEE Applied Power Electronics Conference and Exposition*, pp. 1494–1497
14. R.A. Garcia Mora, *High-Efficiency 3kW Bridgeless Dual-Boost PFC Demo Board*
15. E. Alfawy, R.A. Garcia Mora, *48V lead-acid/Li-ion Battery Charger*

Chapter 13
Optimization of Closed Loop Controlled Charging Time of Li-Ion Battery Using ANFIS

Prakash Bahrani and Naveen Jain

Abstract Increasing applications of Li-ion secondary battery as energy storage device specifically with renewable energy sources (RES), rapid recharging of battery is desired. The most common open loop system of charging batteries is constant current-constant voltage (CC-CV) method, multistage charging, and pulse charging. Temperature rise is an important aspect during the battery charging, which directly affects the life cycle and performance of the battery. The closed loop CT-CV charging takes care the temperature rise condition to improve the life of battery. İn this paper, an existing closed loop control of battery charging is taken as test system, and charging time optimization is done using the ANFIS technology. A rigorous comparative analysis of CC-CV, CT-CV based on PID with feedforward current and ANFIS trained fuzzy logic control simulation is done with models developed in MATLAB/SIMULINK environment. The comparative simulation results exhibit that the charging time with ANFIS is optimized as compared to PID based closed loop system.

Keywords ANFIS · Battery charger · Closed loop control · Fuzzy logic control

13.1 Introduction

A Li-ion battery holds the key for future of renewable energy sources (RES), transportation, portable gadgets, microgrid operations, and many more. The Li-ion battery are preferred due to its high energy storage capacity with light weight, no memory effect, and multiple charge–recharge cycles over other secondary batteries. Further, charging and discharging play crucial role in smooth operation of the utility involving the battery. Many methods like current pumped battery charger, phase locked battery charger, CC-CV, and sensor less charging are available for charging the Li-ion battery [1–5]. Most of the charging techniques does not consider temperature and aging factors. Some methods like sinusoidal ripple charge (SRC) and CT-CV includes the

P. Bahrani (✉) · N. Jain
Maharana Pratap University of Agriculture and Technology, Udaipur, India

A. Khosla and M. Aggarwal (eds.), *Smart Structures in Energy Infrastructure*, Studies in Infrastructure and Control,
https://doi.org/10.1007/978-981-16-4744-4_13

temperature effect in the thermal modeling of the battery and reduced the charging time considerably compared to CC-CV charging incorporating closed loop control methods [4, 6–14]. Consideration of state of charge and internal resistance factor during charging must be taken into account for better life of battery [5]. Some meta-heuristic approach like fuzzy logic controller, finite element analysis, artificial neural network, etc., being accurate approach is applied to reduce the charging time, are presented in studies which requires large training data and time with complex structure to design [10, 15]. While using CC-CV method in CC mode, a constant current is supplied to the battery till the battery voltage reaches to predefined level, and after this point, CV mode comes into action to avoid excessive voltage stress on battery. Depending upon available amount of current through supply during CC mode, charging time can range between 75 and 125 min. Most of CC-CV charger adopts temperature safety and cut-off the charger once the set limit of temperature reaches, which is not exactly is the requirement [14]. In CT-CV technique, a high amount of current is sent to charge the battery during CC mode with set limit of temperature for fast charging and reducing the time during this mode. The amount of charging current is controlled through some closed loop control action and time varying input current.

The closed loop control system data can be more vigorously trained with some heuristics method, which provides better relation between input and output side of the close loop system. Various artificial intelligence (AI) approaches like fuzzy logic and biological swarm optimization are available to solve the problem in different areas of control [13, 16–18]. Artificial neural network (ANN)-based system is one of the emerging control system that develops a modeling by training the available data set from the problem. A comprehensive study on such practice using adaptive network fuzzy inference system (ANFIS) provides a much better technique to obtain the solution of problem with existing method of closed loop system [15]. Many closed loop system has performed well with use of ANFIS technique comparative to some ANN based methods, which encourages the researchers to apply this technique in most possible areas [19–21].

Aim of this study is to use the control data available from closed loop CT-CV technique using conventional PID controller [14] as input to charging current, and to develop a suitable and more efficient control charging model using ANFIS training tool. For better analysis, the study also covers the conventional charging method CC-CV on 4.2 V Li-ion battery system and compares the results of all three methods. The proposed study goes well with the application in areas where limited power is available and consistency with reliability of supply is more required in utility. Reduction of charging time reduces the cost of infrastructure and increases life of battery while considering temperature rise as major concern.

In this paper, a thermal charging model of Li-ion battery is developed in MATLAB/SIMULINK environment to study the CC-CV and CT-CV charging method. The test system under consideration covers temperature control and three different initial state of charge (SoC) of the battery. The ANFIS tool is used to train the data available from closed loop control system in CT-CV method and a suitable fuzzy logic controller is proposed in place of conventional PID controller. Three different initial SoC level namely 0, 20, and 50% are presented in study for better comparison of all the methods.

13.2 Li-Ion Battery Charging Model

With application of control and design methods, the charging time of the battery can be reduced considerably. A proper mathematical modeling for simulation models is presented in this section.

13.2.1 Principle of Li-Ion Battery Charging

Li-ion battery charging can be done in different ways, including pulse charging, multi-stage charging, trickle charge, and boost charging. Boost charging method adopts a constant charging current for a time when the battery gets the desired voltage level, and then, the fixed or constant current is reduced to a small amount of current which then charge the battery till 100% SoC. This method is also known as CC-CV method. CC-CV and CT-CV methods of charging are presented as below.

Constant Current-Constant Voltage (CC-CV) charging

The most common method of battery charging is CC-CV method in which a constant amount of current is supplied to the battery during CC mode till the battery achieves a predecided voltage level. After the desired voltage is reached, the charging current reduced drastically and CV mode is applied till the battery gets fully charged. In the CV mode, charging time is increased considerably and temperature rise aspect is not taken care, during the CC mode temperature rises up to 4–8 °C and may rise more depending upon the charging current for specific application. Application of various optimization techniques may work well with battery parameters but the main objective of cell voltage and cell temperature may change during the optimization cycle [14].

Constant Temperature-Constant Voltage (CT-CV) charging

This is also a closed loop control charging method which provides feedback to the input charging current based on voltage level and temperature conditions of cell and surrounding. In CT mode, temperature rise is considered as from CC charging mode, and taken into account an amount of charging current can be raised or lowered as per temperature feedback of closed loop. With suitable compromise in life cycle of the cell, temperature limit can be raised for even faster charging.

Equivalent Circuit Diagram of Li-Ion Battery

A Li-ion battery can be electrically considered as large capacitor which charge and discharge repeatedly and absorb or supplies electrical power. As shown in Fig. 13.1,

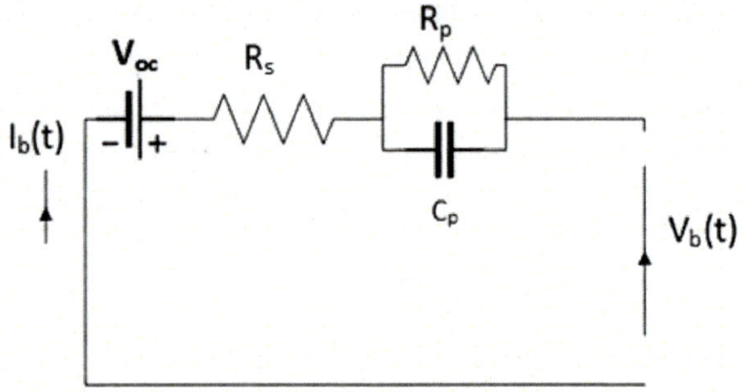

Fig. 13.1 Li-ion battery equivalent circuit diagram [22]

the battery equivalent circuit consists of a series resistance R_s which changes with battery temperature and internal resistance R_p due to electrochemical process.

where

V_{oc} Open circuit battery voltage
$V_b(t)$ Battery terminal voltage
$I_b(t)$ Battery charging current
R_s Battery series resistance
R_p Battery internal resistance
C_p Capacitor consider as battery.

13.2.2 *Mathematical Model Li-Ion Battery Charging*

The battery equivalent model development requires information of the battery voltage, current, SoC, and temperature limit. The effect of temperature is discussed for development of simulation model. The battery model equations considering temperature effect are presented hereunder.

The charging model ($i^* < 0$)

$$f_{ch}\left(it, i^*, T_c, T_{am}\right) = E_0(T_c) - K(T_c)\frac{Q(T_{am})}{0.1Q(T_{am}) + it}i^* - K(T_c)\frac{Q(T_{am})}{Q(T_{am}) - it}it + a\exp(-b.it) - C.it \tag{13.1}$$

$$V_{batt}(T_c) = f_{ch}\left(it, i^*, T_c, T_{am}\right) - R(T_c)i \tag{13.2}$$

where

i^* Low frequency current, A
Q Maximum Capacity, Ah
i Battery current, A

i_t Current drawn from the battery, A
E_{batt} Nonlinear battery voltage, V
E_0 Constant Voltage, V
Exp Exponential zone values
K Polarization constant.
a Voltage during exponential, V
b Capacity during exponential, Ah
T_{ref} Atmospheric temperature, °C
T_c Temperature of cell, °C
T_{am} Ambient Temperature, °C.
The cell temperature can be represented by the following equation

$$T_c(t) = L^{-1}\left(\frac{P_L R_t + T_{am}}{1 + s \cdot t_c}\right) \tag{13.3}$$

where
R_t Resistance based on temperature, Ω
t_c Time constant based on temperature condition.

Thermal model of Li-Ion battery

The surface temperature of battery rises during the charging and discharging process power loss. The battery thermal model can be expressed as equation below.

$$C_i \frac{dT_i}{dt} = P_L - \frac{T_i - T}{R_{is}} \tag{13.4}$$

$$C_s \frac{dT_s}{dt} = \frac{T_i - T}{R_{is}} - \frac{T - T_{am}}{R_{sa}} \tag{13.5}$$

where
C_i Cell capacity based on internal heat
T_i Internal temperature of cell, °C
T Temperature of Cell, °C
C_s Cell capacity based on surface heat
T_{am} Ambient temperature of Cell, °C
R_{is} Cell Thermal resistance in between internal and surface, Ω
R_{sa} Cell resistance between surface and ambient thermal condition, Ω
SoC of the battery can be expressed as

$$\text{SOC}(t) = \frac{Q(t)}{Q_n} \tag{13.6}$$

where $Q(t)$ is current-state capacity, Ah and Q_n is Nominal or rated capacity, Ah.

Description of Test Model for CT-CV charging

In CC-CV method, the rise of temperature during CC mode is observed at the end of the CC mode, and temperature remains in lower level during most of CC mode. This allows us to be able to supply higher current in that region of temperature limit. Hence, the initial current can be taken much higher than 2C as given in CC mode. The charging current must be decreased as the CC phase is ending, and the battery voltage level reaches to desired level. An exponentially decaying current function serves the purpose of variable current profile for such an arrangement during CC mode and with the temperature limit set for charging the setup is addressed as constant temperature (CT) mode in place of CC mode [14]. The exponentially current function given by expression below.

$$I_{\text{ff}} = \left\{ \begin{array}{cc} 2C: & 0 \le t < t_{\text{pk}} \\ C\left(1 + \mathrm{e}^{-\frac{(t-t_{\text{pk}})}{\tau}}\right): & t_{\text{pk}} \le t < t_{\text{cv}} \end{array} \right\} \tag{13.7}$$

where

t_{pk} Time taken by current to reach its peak value, seconds
t_{cv} Time for CV mode, seconds
Ţ Time constant
I_{ff} Exponential feed forward current.

The temperature is kept within the limit to the cell's permissible temperature, and a PID controller is adopted to provide temperature feedback to the input feedforward current. The basic equation of PID control system is expressed in equation below.

$$u(t) = K_{\text{p}}e(t) + K_{\text{i}}\int_{0}^{t} e(\tau)\mathrm{d}\tau + K_{\text{d}}\frac{\mathrm{d}e(t)}{\mathrm{d}t} \tag{13.8}$$

13.3 Adaptive Neuro-Fuzzy Inference System (ANFIS) Training

Adaptive neuro-fuzzy inference system (ANFIS) is the technique or learning machine, which explores the approximation using neural network to provide most suitable parameters to the fuzzy inference system. It was developed in early 1990's [23], and it utilizes hybrid learning methods to provide solutions for ill-defined models and nonlinear complex systems. The ANFIS exploits both the neural network and fuzzy logic system to optimize the input and output data sets. The basic structure of ANFIS is shown in Fig. 13.2 as a poly-layer feedforward network, which has nodes in each layer. The structure consists of two inputs, one output, and nine rules as shown in figure of five layer ANFIS structure. In this structure, Layer-1 calculates the membership function from input signal, Layer-2 defines the rule for each

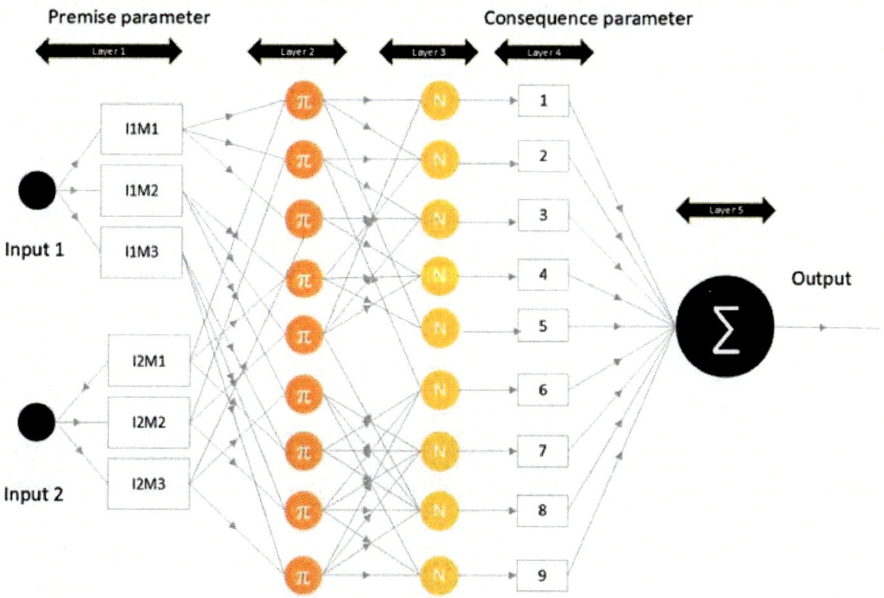

Fig. 13.2 Architecture of ANFIS with 2 input 1 output and 9 rules

membership function, Layer-3 calculates the scaling value of each output membership function, and Layer-4 calculates the value of each output. Finally, Layer-5 gives overall output of the ANFIS.

For the accuracy analysis, the data are trained and tested as per Table 13.1 as listed. The training of data is done in two ratio, namely 70–30% and 80–20% for each initial SoC condition. The ANFIS parameter for training are listed in Table 13.1. The dataset is taken from stored data of PID based CT-CV model (Table 13.2).

Table 13.1 ANFIS training parameter

Parameter	Signal from dataset	MF	Fuzzy rulebase	Training algorithm	Epoch	Dataset
Premise parameter	Error in temperature	3	Gbell linear values Sugeno method	MPSO	3 (Max 25) Training stopped at epoch 2	Available from CT-CV PID
	Change in error	3				train–test–check ratio 80–10–10 and 70–20–10
Consequent parameter	Current signal	9	–	RLSE		

Table 13.2 ANFIS training error analysis

Train–test data		70–30%	80–20%
SoC	0%	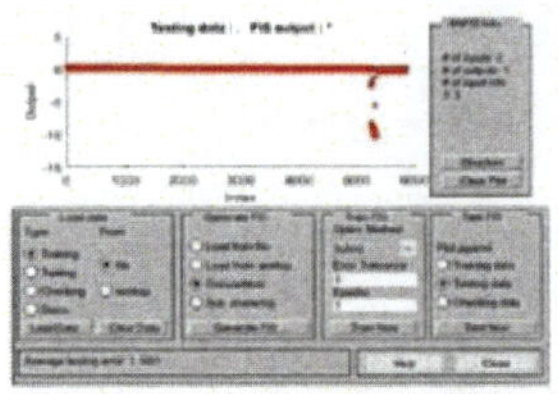	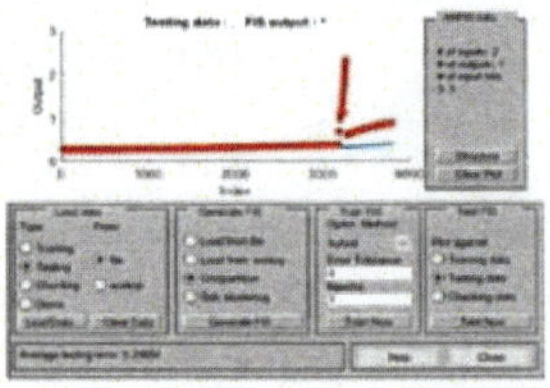
	20%	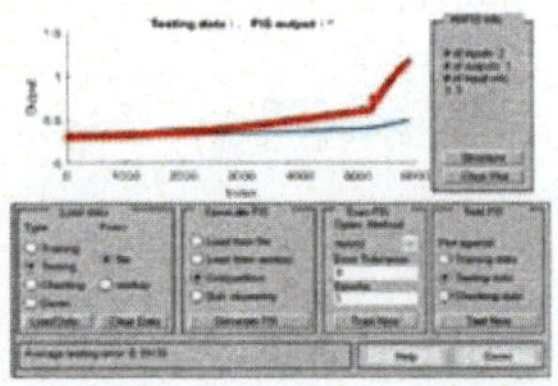	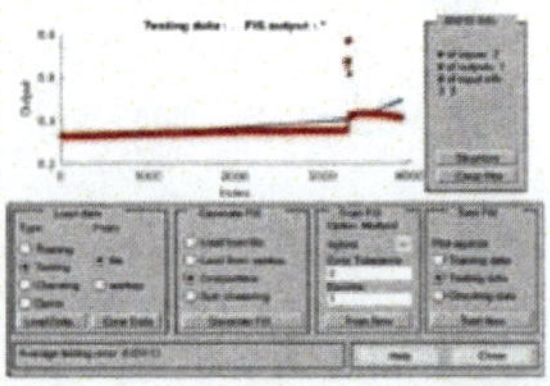
	50%	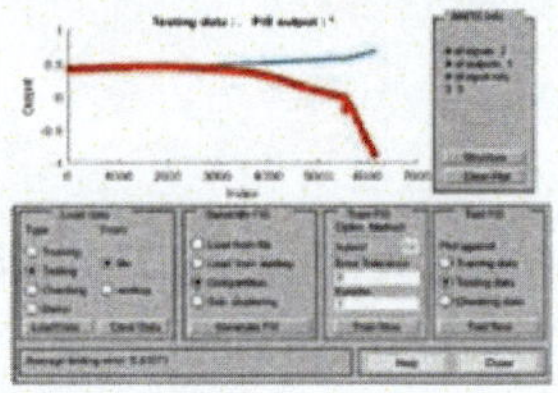	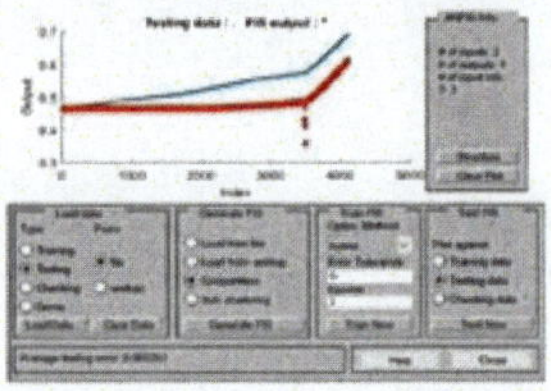

13.4 Proposed Algorithm, Simulation Result and Discussion

The PID block of the proposed test system is replaced with suitable fuzzy inference trained by ANFIS. The simulation is performed in MATLAB/SIMULINK R2018a environment to obtain the results. All the three modes, namely CC-CV, CT-CV with PID and CT-CV with ANFIS data are executed with 0, 20, and 50% initial SoC of Li-ion 4.2 V battery. Data training in ANFIS is done as per the details illustrated in Table 13.1, and the train–test–check ratio is considered for two conditions. The flowchart for the test system is presented as below in Fig. 13.3.

Comparative simulation results are given in Tables 13.3, 13.4, and 13.5 for all the test conditions with initial SoC of 0, 20, and 50%.

The simulation results presented in Table 13.4 demonstrate the comparative study results for SoC.

The simulation results presented in Table 13.5 demonstrate the comparative study results for temperature rise.

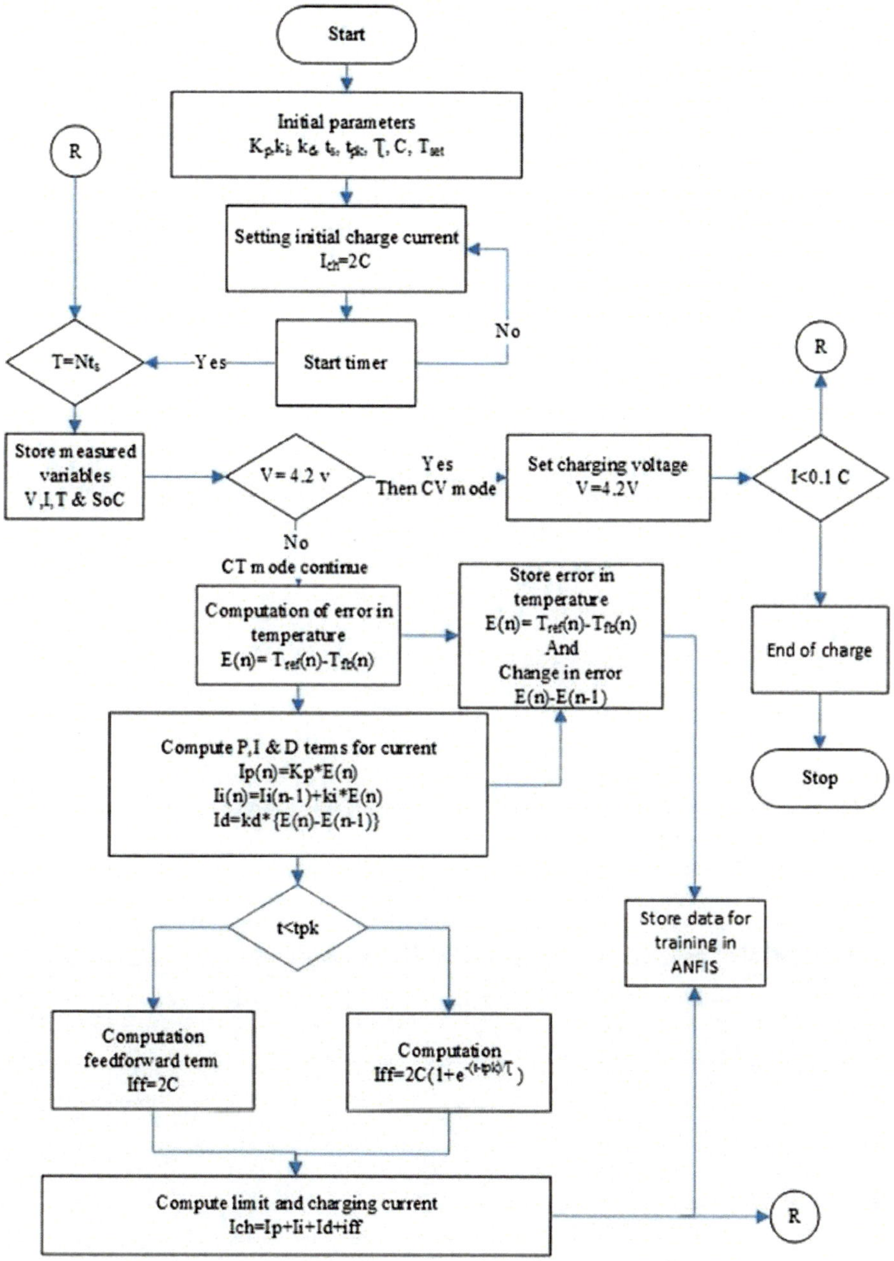

Fig. 13.3 Flowchart for proposed CT-CV charging using ANFIS training

Table 13.3 Fuzzy inference rule base

Trained data 80–20%		
0% initial SoC	20% initial SoC	50% initial SoC
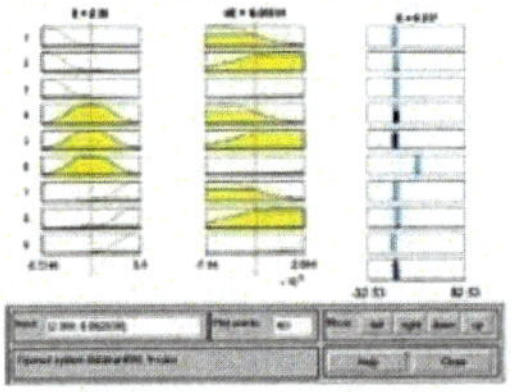	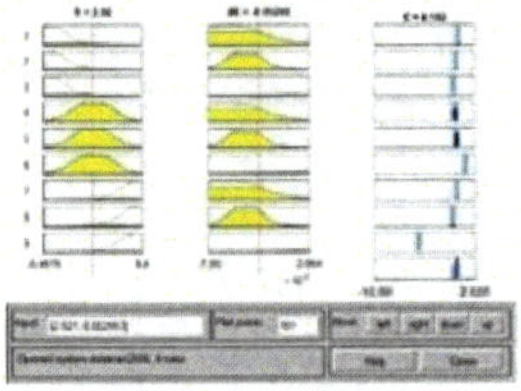	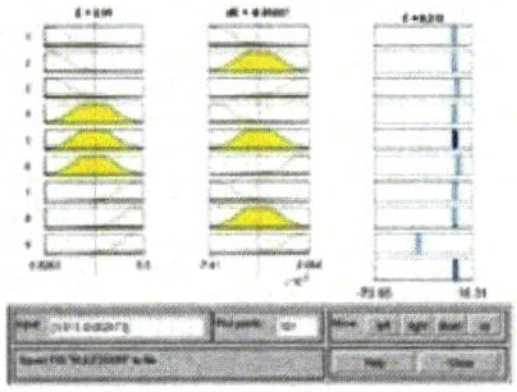

Table 13.4 Comparative result analysis for SoC of Li-ion battery

Time to reach 100% SoC		
0% initial SoC	20% initial SoC	50% initial SoC
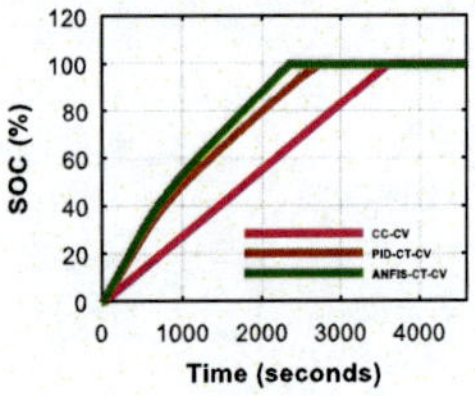	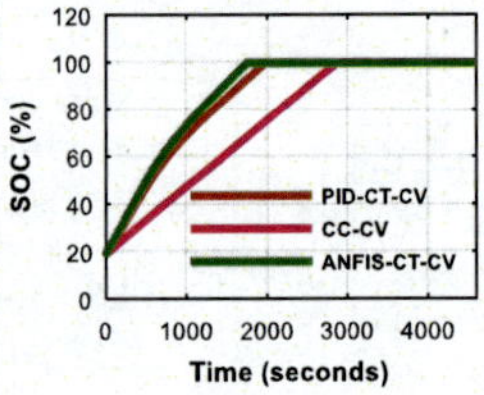	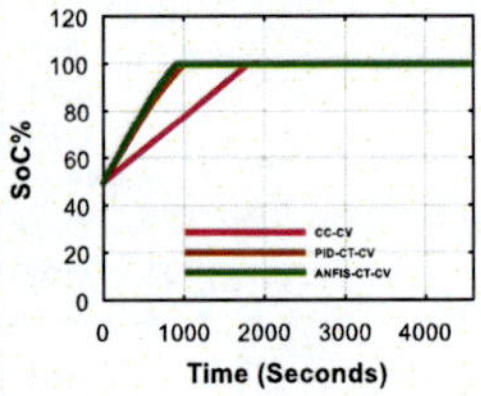

Table 13.5 Comparative result analysis for temperature rise of Li-ion battery

Temperature rise during charging		
0% initial SoC	20% initial SoC	50% initial SoC
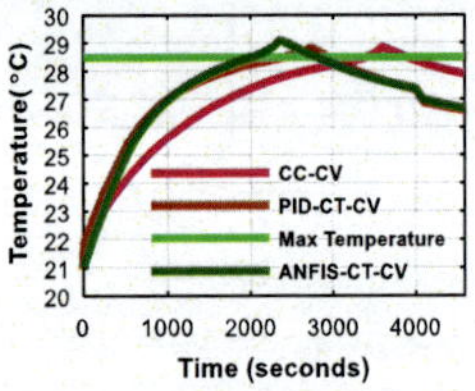	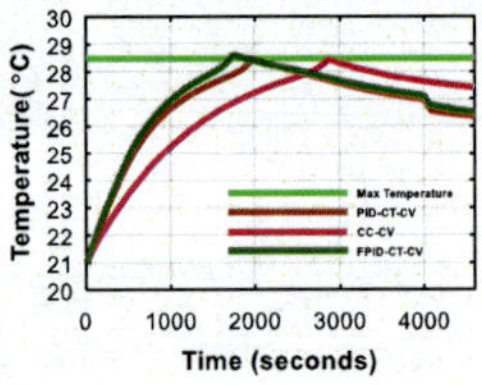	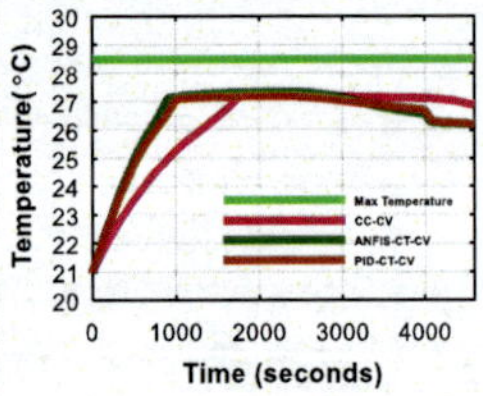

13.5 Conclusion

In this paper, an optimized method for closed loop control charging system of Li-Ion 4.2 V battery is presented. Batch training of the input output data obtained from conventional closed loop control is used in ANFIS with hybrid training at 70:30 and 80:20 ratio to train. The fuzzy inference, thus, obtained via ANFIS is applied to the test system to check system performance in MATLAB simulation environment. The results of the simulation study shows that the trained data through ANFIS provides better optimization. In case of initial SoC 0%, 20%, and 50%, the ANFIS model is found to be faster by 5%, 3.3%, and 6%, respectively.

References

1. L. Chen, J.J. Chen, C. Neng-Yi, H. Gia-Yo, Current-pumped battery charger. IEEE Trans. Ind. Elect. **6**, 2482–2488 (2008)
2. L.R. Chen, R.H. Chaoming, C.S. Liu, A design of a grey-predicted li-ion battery charge system. IEEE Trans. Ind. Elect. **10**, 3692–3701 (2008)
3. L.R. Chen, C.S. Liu, J. Chen, Improving phase-locked battery charger speed by using resistance-compensated technique. IEEE Trans. Ind. Elect. **4**, 1205–1211 (2009)
4. R.C. Cope, Y. Podrazhansky, The art of battery charging, in ed. By H.A. Frank, E.T. Seo (eds.) *Proceedings of 14th Annual Battery Conference on Applied Advance* (Ready Reproductions Inc., California, Long Beach, 1999), pp. 233–235
5. K. Chung, S. Hong, O. Kwon, A fast and compact charger for li-ion battery using successive built-in resistance detection. IEEE Trans. Circt. Syst. **2**, 161–165 (2017)
6. L.R. Chen, S.L. Wu, D.T. Shieh, T.R. Chen, Sinusoidal-ripple-current charging strategy and optimal charging frequency study for Li-ion batteries. IEEE Trans. Ind. Elect. **1**, 88–97 (2013)
7. D. Karaboga, E. Kaya, Adaptive network based fuzzy inference system (ANFIS) training approaches: a comprehensive survey. Artif. Intell. Rev. **52** (2020)
8. H.T. Lin, T.J. Liang, S.M. Chen, Estimation of battery state of health using probabilistic neural network. IEEE Trans. Ind. Inform. **2**, 679–685 (2013)
9. J. Liu, D. Gao, J. Cao, Study on the effects of temperature on LiFeP04 battery life, in *IEEE Vehicle Power and Propulsion Conference*, vol. 8 (IEEE Press, Seoul, South Korea, 2012), pp. 9–12
10. N. Liu, Q. Chen, X. Lu, J. Liu, J. Zhang, A charging strategy for PV-based battery switch stations considering service availability and self-consumption of PV energy. IEEE Trans. Ind. Elect. **62**, 4878–4889 (2015)
11. F. Liu, H. Wang, Q. Shi, H. Wang, M. Zhang, H. Zhao, Comparison of an ANFIS and fuzzy pid control model for performance in a two-axis inertial stabilized platform. IEEE Access **5**, 12951–12962 (2017)
12. A.E. Mejdoubi, A. Oukaour, H. Chaoui, H. Gualous, J. Sabor, Y. Slamani, State-of-charge and state-of-health lithium-ion batteries diagnosis according to surface temperature variation. IEEE Trans. Ind. Elect. **4**, 2391–2402 (2016)
13. Z. Miao, L. Xu, V.R. Disfani, L. Fan, An SOC-based battery management systems for microgrids. IEEE Trans. Smart Grid **5**, 966–973 (2014)
14. N.S. Motapon, L.B. Alexandre, L.A. Dessaint, H. Fortin-Blanchette, K. Al-Haddad, A generic electro-thermal li-ion battery model for rapid evaluation of cell temperature temporal Evolution. IEEE Trans. Ind. Elect. **2**, 151–160 (2016)
15. M.S. Golsorkhi, D.D.C. Lu, A decentralized control method for islanded microgrids under unbalanced conditions. IEEE Trans. PowerDel **31**, 1–9 (2016)
16. L. Patnaik, A.V.J.S. Praneeth, S.S. Williamson, A closed-loop constant-temperature constant-voltage charging technique to reduce charge time of lithium-ion batteries. IEEE Trans. Ind. Elect. **2**, 1059–1067 (2018)
17. T.L. Vandoorn, B. Meersman, L. Degroote, B. Renders, L. Vandevelde, A control strategy for islanded microgrids with DC-Link voltage control. IEEE Trans. Power Del. **26**, 703–713 (2011)
18. T. Wang, D. O'Neill, H. Kamath, Dynamic control andoptimization of distributed energy resources in a microgrid. IEEE Trans. Smart Grid **6**, 1–11 (2015)
19. U.K. Das, S. Samantaray, D.K. Ghose, P. Roy, Estimation of aquifer potential using BPNN, RBFN, RNN, and ANFIS, ed. by S. Satapathy, V. Bhateja, S. Das (eds.) *Smart Intelligent Computing and Applications* SIST, vol. 105 (Springer, Singapore, 2019)

20. I. Entehaj, H. Bonakdari, S.M. Sadeg, Design of a hybrid ANFIS-pso model to estimate sediment trasport in open channels. Iran J. Sci. Technol. Trans. Civ. Eng. **43**, 851–857 (2019)
21. L. Xiaonan, S. Kai, J.M. Guerrero, J.C. Vasquez, H. Lipei, Double quadrant state-of-charge-based droop control method for distributed energy storage systems in autonomous DC microgrids. IEEE Trans. Smart Grid **6**, 147–157 (2015)
22. L. Gao, S. Liu, R.A. Dougal, Dynamic li-ion battery model for system simulation. IEEE Trans. Compon. Packag. Technol. **25**, 495–505 (2002)
23. B. Peng, S. Wang, Y. Liu, Y. Syun, Huang, A li-ion battery charger based on remaining capacity with fuzzy temperature control. IEEE Trans. Ind. Elect. **5**, 112–122 (2016)

Chapter 14
Electric Vehicle Reliability Assessment Based on Fault Tree Analysis

Kailash Rana and Dheeraj Joshi

Abstract With increasing environmental concerns, electric vehicles (EV) are gaining more and more attention. The main components of EV are composed of power electronic devices, and conducting reliability analysis has an important impact on its safety. To do so, the reliability of EV is analyzed based on a fault tree (FT). For reliability analysis, firstly the theoretical failure rates of individual subassemblies are considered which is composed of a DC-DC (MIMO), inverter, and a motor drive. Then the reliability of the entire system is evaluated. The results of the reliability calculations showed the most prone to failure components.

Keywords Electric vehicle · Reliability · Fault-Tree

14.1 Introduction

Internal combustion engine (ICE) vehicles propelled by fossil fuels largely dominate the automotive sector. This mature technology offers high power along with fast refueling capability which makes fuel-based vehicles cost effective, robust, and functional for the consumers. However, major concerns have been raised regarding the sustainability of non-renewable energies and their adverse impact on the environment and human health. In 2016, the world devoured more than 13.3 trillion tons of energy of which the oil consumption accounted for 33%. The transport sector consumed about 20% of the total energy demand, and oil is its primary fuel source with 94%. Energy forecast shows that world oil consumption in transportation will rise by 28.7% by 2040 [1]. Hence, energy diversification is essential to reduce fossil fuel usage and avoid dependence on a sole energy source.

Thus, there has been a rising interest to substitute conventional combustion engines with electric ones in order to minimize the effects of the issues described

K. Rana (✉) · D. Joshi
Delhi Technological University, Delhi, India

D. Joshi
e-mail: joshidheeraj@dtu.ac.in

A. Khosla and M. Aggarwal (eds.), *Smart Structures in Energy Infrastructure*,
Studies in Infrastructure and Control,
https://doi.org/10.1007/978-981-16-4744-4_14

above. Electric vehicles (EVs) have been widely attracted many researchers, as a crucial and non-polluting transportation tool [2]. It is deemed that pure electric vehicles are expected to supersede traditional combustion vehicles at some stage in future. This implies in the near future many EVs will be running on the road.

Reliability evaluation of electric units such as the battery, converter, and electric motor plays a critical role in the lifetime performance of the EV system [3]. Power electronic (PE) converters and motor are the key components of EVs. Its reliability has an important impact on the safety of the system [4].

So far, various techniques have been discussed regarding reliability evaluation and have shown that failure of single component can cause failure [5, 6]. An approach has been proposed in [7] based on the part-count model of interleaved boost converter, which shows individual component failure and degradation study. Marzieh Piri et al. present Markov model-based reliability analysis of PV system and considered different PE components [8]. In [9], physics of failure (PoF)-based analysis is done on a wind power converter to perform reliability-oriented design of PE system.

Consequently, several reliability-oriented issues are investigated on EV systems which are mainly concentrated on how to enhance performance of the system based on fault tolerance and control [10]. It can be seen from the above literature that the current research was mainly focused on power electronics converter reliability, reliability problems existing in motor drive systems, reliability issues in the photovoltaic system and wind systems, and the issues related to reliability analysis of the whole EV system are rarely discussed. This may lead to inaccurate research conclusions owing to the fact that the power electronic converters and motor drives are logically integrated. A combined reliability assessment of DC-DC converter, inverter, and the motor will provide a more accurate prediction of the whole EV. For this reason, this paper's research will focus on evaluating the reliability of the whole EV system. The main parts of pure EV are shown in Fig. 14.1.

The fault tree approach, a powerful technique for evaluating reliability and safety evaluation, has been widely employed to evaluate the reliability concerns in different systems. For example, in [11], fault tree method is used to evaluate potential failure

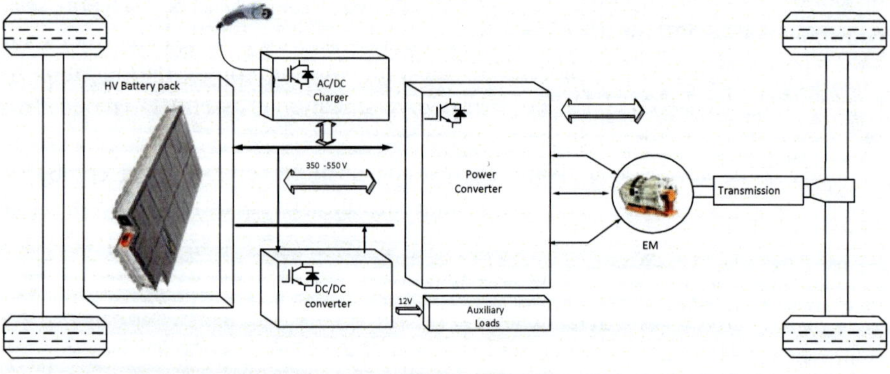

Fig. 14.1 Electric vehicle system

of the wind turbine; for qualitative evaluation of complex system containing multi-component systems, the fault tree method is used in [12]; for accessing reliability of fault-tolerant control system, dynamic fault tree model was developed in [13]. In this paper, reliability assessment of EV system is done based on fault tree analysis due its effective and extensive use for reliability analysis. The remaining sections of the paper are structured as below: In Sect. 14.2, reliability issues in power electronics will be briefly discussed so as the reader can have a basic understanding of it. In Sect. 14.3, reliability-oriented design of EV system is discussed, and based on fault tree method, reliability is analyzed. Finally, the paper will be concluded in Sect. 14.4.

14.2 Reliability Issues in Power Electronics

The reliability of a component is defined as the probability that a component will perform a required work without any failure under the stated situation for a stipulated period of time [14]. The purpose of reliability engineering is to prevent the creation of any failure. Today, the industry is in the midst of a paradigm shift toward more dependable power electronic systems. Power electronic system comprises of many components which experience thermal, chemical, electrical, mechanical stress and degrade gradually which leads to complete failure [15, 16] (Fig. 14.2).

Reliability of power electronics encompasses many disciplines, which include component engineering, electrical engineering, material science engineering, production engineering, mechanical engineering, test engineering, reliability engineering, and failure investigations. Each of these needs to operate as a single system to achieve the desired goal based on industry requirements (Table 14.1).

The bath curve in Fig. 14.3, shows equipment/parts in all phases of life which includes the infant period, useful lifetime, and wear-out period. The failures during the infant period are due to material defects or design defects. Wear-out failures are due to material fatigue or depletion after a long period of use. Useful lifetime failures are the failures that occur during the normal operational period due to stresses.

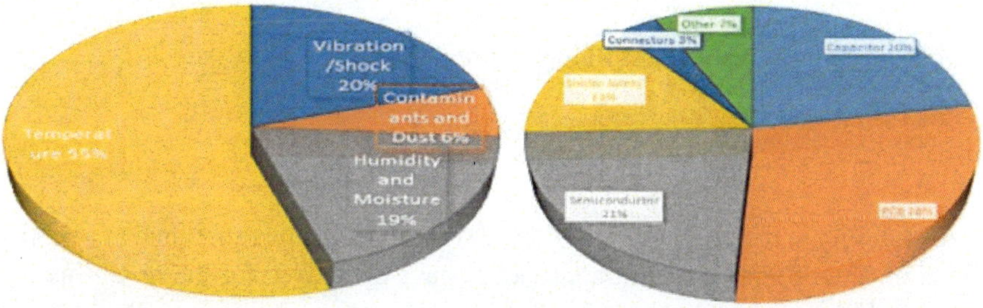

(a) Stresses distribution (b) Failure distribution

Fig. 14.2 Stress and failure distribution in power electronics systems [10]

Table 14.1 Lifetime target in different PE systems [4]

Applications	Typical design target of lifetime
Photovoltaic plants	Five to thirty years (12 h per day)
Automotive	Fifteen years (300,000 km, operating hours 10,000)
Industry motor drives	Five to twenty years (full load for 60,000 h)
Wind turbines	Twenty years (operation per day—24 h)
Aircraft	Twenty-four years (flight operation—100,000 h)
Railway	Twenty to thirty years (operation per day—10 h)

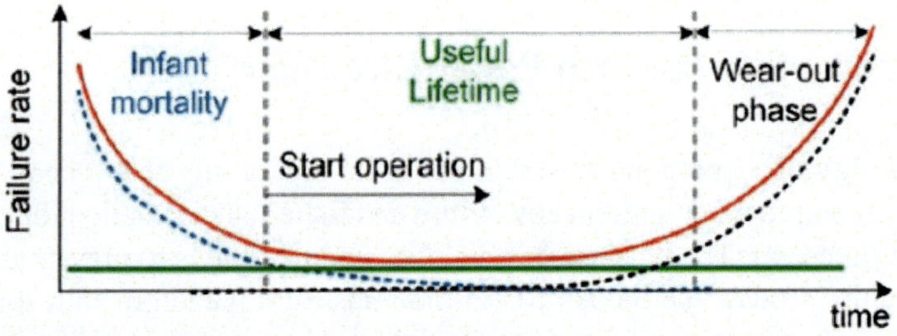

Fig. 14.3 Typical bath curve describing failure rate of item

The reliability function $R(t)$ is shown in Eq. (14.1), where $\lambda(t)$ represents the failure rate [5].

$$R(t) = e^{-\lambda t} \tag{14.1}$$

Mean time to failure (MTTF) is time period unit system fails, and it is described by Eq. (14.2).

$$\text{MTTF} = \int_{0}^{+\infty} R(t)\text{d}t \tag{14.2}$$

For constant failure rate, MTTF is expressed as

$$\text{MTTF} = \frac{1}{\lambda} \tag{14.3}$$

Availability is the probability that the device/unit will be functional at a particular period of time. So far different approaches have been used for the reliability evaluation [5]. Most common approaches are MIL-HDBK-217, Siemens SN29500, Bellcore/Telcordia (electronic handbook for reliability), NSWC-06/LE10 (Naval reliability handbook), China 299B (Chinese reliability program) RDF 2000 (French reliability handbook), and several others. These handbooks for reliability prediction do not give any clarity and command over the real sources of failure since the reason behind any failure which effect reliability is not taken [17]. System or subsystem level reliability model can be categorized into part-count, combinational, and state-space model [18].

14.3 Reliability-Oriented Design of EV System Based on Fault Tree

For reliability analysis of EV system, there is a need to consider the potential reasons which may give rise to failure and building reliability model of the system. Most common techniques for constructing reliability model are reliability block diagram (RBD), Markov model, and fault tree model. Fault tree method is based on finding individual component probability and logical relationship between different components. With fault tree (FT) analysis, we can identify cause of failure and estimate the reliability details of complete structure. Since fault tree method has been applied to numerous systems to the reliability as discussed earlier, it is taken to analyze the reliability of EV system. An EV system consists of several components, and the core components consist of DC-DC converter, inverter, and drive motor. For the reliability study of the system, considered DC-DC converter based on multi-input multi-output configuration [6, 19], conventional three leg inverter, and motor drive.

In order to determine the reliability of subsystems in the EV system, fault tree of complete EV system is presented in Fig. 14.4. As in figure, the top-event E1 is EV system failure and F1 to F3 are the intermediate events, which are the logic gate events. The basic events from Fc1 to Fc3 are the failure events of MIMO converter units, Fi1 to Fi3 are the failure events for inverter units, and Fmx1 to Fmx9 are the failure events for motor units.

All events of power converter and motor units are explained in Tables 14.2 and 14.3, respectively.

From Fig. 14.4, it is observed that the reliability of the EV system depends on the MIMO converter reliability, inverter reliability, and motor reliability. For investigating the reliability of EV system, it is essential to assess the failure rate of the associated components and their respective reliability probability functions. Basic event failure rates are estimated using the MIL-HDBK handbook and the US Navy's reliability handbook [6, 20]. The failure rate of diode capacitor, switch, and inductor can be calculated, respectively, by [21, 22].

$$\lambda_{P(S)} - \lambda_{b(s)} \times \pi_Q \times \pi_A \times \pi_E \times \pi_T \tag{14.4}$$

$$\lambda_{P(D)} = \lambda_{b(D)} \times \pi_Q \times \pi_E \times \pi_T \times \pi_S \times \pi_C \tag{14.5}$$

$$\lambda_{P(L)} = \lambda_{b(L)} \times \pi_Q \times \pi_C \times \pi_E \tag{14.6}$$

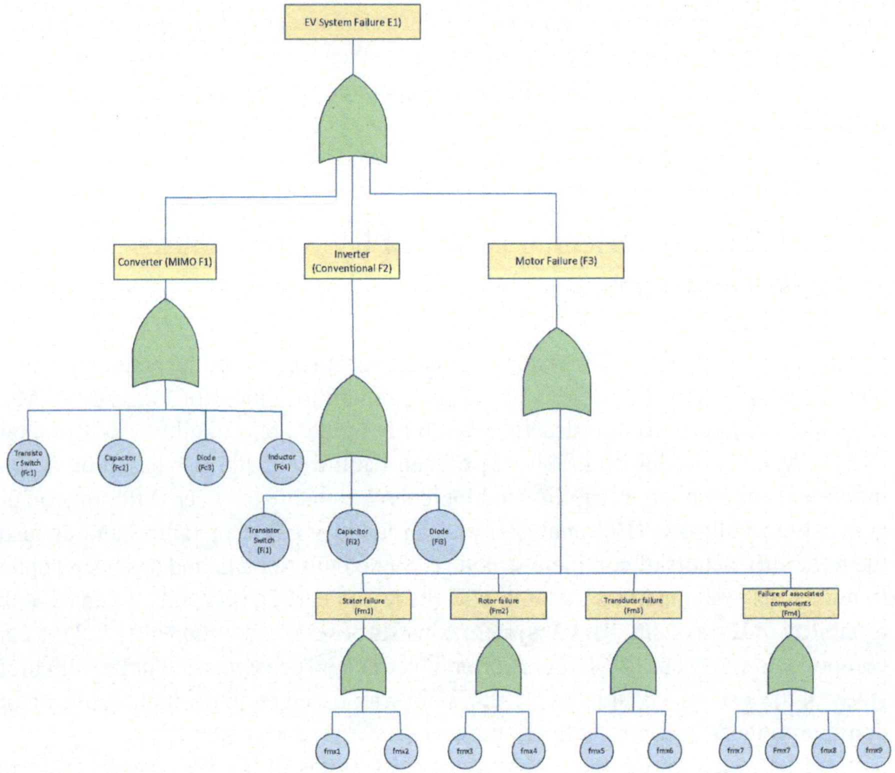

Fig. 14.4 Fault tree of EV system

Table 14.2 Failure rates of power converter

Intermediate-event	Code	Failure rate	Basic event	Code	Failure rate (FPMH-Total)
MIMO converter failure	F_1	λ_{F1}=43.706	Failure of switch	Fc1	λ_{Fc1}=41.86
			Failure of capacitor	Fc2	λ_{Fc2}=1.14
			Failure of diode	Fc3	λ_{Fc3}=0.68
			Failure of inductor	Fc4	λ_{Fc4}=0.0268
Inverter failure	F_2	λ_{F2}=84.219	Failure of switch	Fi1	λ_{Fi1}=82.92
			Failure of capacitor	Fi2	λ_{Fi2}=0.279
			Failure of diode	Fi3	λ_{Fi3}=1.02

$$\lambda_{P(C)} = \lambda_{b(C)} \times \pi_Q \times \pi_{CV} \times \pi_E \tag{14.7}$$

The reliability of MIMO converter depends on its associated components reliability. Thus, the failure rate of MIMO converter is total sum of all components rate of failure, i.e.,

Table 14.3 Failure rates of drive motor components

Intermediate event	Code	Failure rate	Basic event	Code	Failure rate (FPMH)
Stator failure	F_{m1}	λ_{Fm1}=0.2523	Failure of stator winding	F_{mx1}	λ_{Fmx1}=0.2520
			Failure of stator core	F_{mx2}	λ_{Fmx2}=0.0003
Rotor failure	F_{m2}	λ_{Fm2}=0.2998	Failure of rotor armature winding	F_{mx3}	λ_{Fmx3}=0.2772
			Failure of rotor shaft	F_{mx4}	λ_{Fmx4}=0.0226
Transducer failure	F_{m3}	λ_{Fm3}=0.2580	Failure of temperature sensor	F_{mx5}	λ_{Fmx5}=0.2195
			Failure of position sensor	F_{mx6}	λ_{Fmx6}=0.0375
Failure of other motor components	F_{m4}	λ_{Fm4}=0.5689	Failure of spline	F_{mx7}	λ_{Fmx7}=0.0385
			Failure of bearing oil seal	F_{mx8}	λ_{Fmx8}=0.4465
			Failure of fastening screw	F_{mx9}	λ_{Fmx9}=0.0003
			Failure of bearing	F_{mx10}	λ_{Fmx10}=0.083

$$\lambda_{F1} = \lambda_{Fc1} + \lambda_{Fc2} + \lambda_{Fc3} + \lambda_{Fc3} \tag{14.8}$$

Similarly, the failure rate of inverter and motor is shown below

$$\lambda_{F2} = \lambda_{Fi1} + \lambda_{Fi2} + \lambda_{Fi3} \tag{14.9}$$

$$\lambda_{F3} = \lambda_{Fm1} + \lambda_{Fm2} + \lambda_{Fm3} + \lambda_{Fm4} \tag{14.10}$$

Therefore, the failure rate of whole system can be written as

$$\lambda_E = \lambda_{F1} + \lambda_{F2} + \lambda_{F3} \tag{14.11}$$

Failure rates of individual components are shown in Tables 14.2 and 14.3. The failure rates of whole system are calculated based on number of associated components. From the above calculated data shown in Tables 14.2 and 14.3, it is found that capacitors and switches are the most vulnerable part in the power converter system and bearing oil seal is the most prone to failure component in drive motor.

Therefore, the MTTF of the electric vehicle system which is expected time until complete system fails is 7733.66 h. The reliability at 10,000 h is 0.27, as represented in green shaded area. The probability of failure, is 0.73, as represented by pink shaded area in Fig. 14.5.

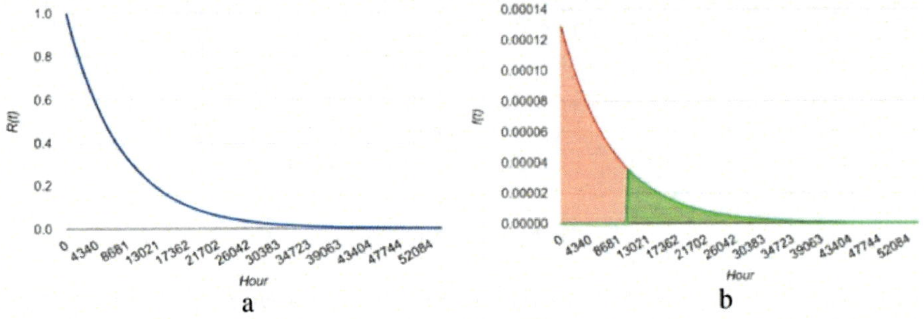

Fig. 14.5 **a** Reliability—*R*(*t*) of EV system. **b** Probability density function—$f(t)$ of EV system at $t = 10{,}000$ h

14.4 Conclusion

The reliability assessment of electric vehicle system has been presented based on fault tree model. It is found that the most prone to failure components of EV system are switches and capacitors. Also, the study reveals that the most vulnerable component of the system so that timely service can be done to avoid complete failure of system.

References

1. T. Muneer, A. Doyle, M.L. Kolhe, *Electric Vehicles: Prospects and Challenges* (Elsevier, 2017)
2. M. Burgess, N. King, M. Harris, E. Lewis, Electric vehicle drivers' reported interactions with the public: driving stereotype change? Transp. Res. Part F: Traffic Psychol. Behav. **17**, 33–44 (2013)
3. B. Wang, G. Tian, Y. Liang, T. Qiang, Reliability modeling and evaluation of electric vehicle motor by using fault tree and extended stochastic petri nets. J. Appl. Math. **2014**, 1–9 (2014)
4. H. Wang et al., Transitioning to physics-of-failure as a reliability driver in power electronics. IEEE J. Emerg. Sel. Top. Power Electron. **2**(1), 97–114 (2014)
5. S. Peyghami, Z. Wang, F. Blaabjerg, A guideline for reliability prediction in power electronic converters. IEEE Trans. Power Electron. **35**(10), 10958–10968 (2020)
6. *Reliability Prediction of Electronic Equipments (MIL-HDBK-217)* (1990)
7. D. Umarani, R. Seyezhai, Reliability assessment of two-phase interleaved boost converter. Life Cycle Reliab. Saf. Eng. **7**, 43–52 (2018)
8. M. Piri, M. Niroomand, R.-A. Hooshmand, A comprehensive reliability assessment of residential photovoltaic systems. J. Renew. Sustain. Energy **7**, 053116 (2015)
9. H. Wang, K. Ma, F. Blaabjerg, Design for reliability of power electronic systems, in *IECON 2012—38th Annual Conference on IEEE Industrial Electronics Society*, Montreal, QC, 2012, pp. 33–44
10. I. Bolvashenkov, J. Kammermann, H.G. Herzog, Research on reliability and fault tolerance of multi-phase traction electric motors based on Markov models for multi-state systems, in *Proceedings of the SPEEDAM*, Aug 2016, pp. 1166–1171
11. J. Kang, L. Sun, C.G. Soares, Fault tree analysis of floating offshore wind turbines. Renew. Energy **133**, 1455–1467 (2019)

12. T.P. KhanhNguyen, J. Beugin, J. Marais, Method for evaluating an extended fault tree to analyses the dependability of complex systems: application to a satellite-based railway system. Reliab. Eng. Syst. Saf. **33**, 300–313 (2014)
13. B. Hu, P. Seiler, Pivotal decomposition for reliability analysis of fault tolerant control systems on unmanned aerial vehicles. Reliab. Eng. Syst. Saf. **140**, 130–141 (2015)
14. P.D. O'Connor, P. O'Connor, A. Kleyner, *Practical Reliability Engineering* (Wiley, Hoboken, NJ, 2012)
15. M. Ciappa, Selected failure mechanisms of modern power modules. Microelectron. Reliab. **42**(4–5), 653–667 (2002)
16. Y. Ong, B. Wang, Survey on reliability of power electronic systems. IEEE Trans. Power Electron. **28**(1), 591–604 (2013)
17. J. Jones, J. Hayes, Estimation of system reliability using a "non-constant failure rate" model. IEEE Trans. Reliab. **50**(3), 286–288 (2001)
18. M.J. Cushing, D.E. Mortin, T.J. Stadterman, A. Malhotra, Comparison of electronics-reliability assessment approaches. IEEE Trans. Reliab. **42**(4), 542–546 (1993)
19. S. Upadhyaya, K. Rana, M. Taneja, D. Joshi, Modelling and control of non-isolated multiport DC/DC converter, in *2020 First IEEE International Conference on Measurement, Instrumentation, Control and Automation (ICMICA)*, Kurukshetra, India, 2020, pp. 1–5
20. *Handbook of Reliability Prediction Procedures for Mechanical Equipment* (Potomac, MD, USA, 2009), pp. 14–29
21. B. Abdi, A.H. Ranjbar, G.B. Gharehpetian, J. Milimonfared, Reliability considerations for parallel performance of semiconductor switches in high-power switching power supplies. IEEE Trans. Ind. Electron. **56**, 2133–2139 (2009)
22. X. Shu, Y. Guo, W. Yang, K. Wei, Y. Zhu, H. Zou, A detailed reliability study of the motor system in pure electric vans by the approach of fault tree analysis. IEEE Access **8**, 5295–5307 (2020)

Chapter 15
Design Analysis of Dimmer Light for Autonomous Vehicles

Abhisheak Mangla, Dhruv Gulati, Nanak Jhamb, and Devendra Vashist

Abstract One of the most innovative features is the invention of an autonomous car. An autonomous car controls the motion, sensor activation and action automatically without any human intervention. Such vehicles ensure high degree of safety, comfort, and ease of driving. Front lights which make one of the important components of such vehicles has to be designed in such a way that it consumes less power and provides a clear vision to occupants. The study aims to develop intelligent operated light dimmer using the concept of Internet of things (IoT). The designed automated light management system automatically manages light intensity and beam settings based on input parameters received from the environment. The system manages the ON/OFF function of lights using an LDR sensor, high/low beams using a front body mounted camera, left/right movement of lights according to steering position and up/down movement of lights according to incline/decline is detected. It makes use of sensors, i.e., BNO055 9-DOF which senses current driving conditions, i.e., incline/decline or turning of the vehicle, BME280 which indicates temperature, humidity, and altimeter sensing that is utilized for calculation of loss in visibility on the road due to fog, smog, rainfall, etc. HB100 sensor is microwave radar that can detect objects up to 16 m helps in choosing beaming (high or low) while ALS-PT19 detects amount of ambient light in the environment to decide ON/OFF or level of headlight brightness. The designed analysis provided satisfactory results and can be successfully fitted in autonomous vehicles.

Keywords Autonomous · Vehicles · Dimmer light · Sensor · Safety

15.1 Introduction

Self-operating light dimming (S.O.L.D) uses the concept of Internet of things (IoT) [1–4]. Automated light management system for vehicles which automatically

A. Mangla · D. Gulati · N. Jhamb · D. Vashist (✉)
Automobile Engineering Department, Manav Rachna International Institute of Research and Studies, Faridabad, Haryana 121010, India
e-mail: devendra.fet@mriu.edu.in

A. Khosla and M. Aggarwal (eds.), *Smart Structures in Energy Infrastructure*,
Studies in Infrastructure and Control,
https://doi.org/10.1007/978-981-16-4744-4_15

manages light intensity as well as beaming settings based on various parameters collected from the environment is created for the vehicle. This system automatically turns ON/OFF the vehicle headlights according to the environment variables, and depending on the necessity of the operation, the specific action performs actively or passively. Apart from controlling the ON/OFF function of the headlight, supported lights can also be dimmed to better suit road and traffic conditions without affecting the fellow drivers. For the same reason, beaming state of the light (high or low) is also managed by the system according to the needs.

15.1.1 Available Technologies

The different available technologies for managing dimmer light for autonomous vehicles are

a. ON/OFF function of lights using an LDR sensor.
b. High/low beams using a front body-mounted camera.
c. Left/right movement of lights according to steering position.
d. Up/down movement of lights according to incline/decline detected.

The existing technology suffers from the following drawbacks

a. High cost because of this only premium cars can afford these features.
b. Some of the features of these technologies has become obsolete as now improvement was made since a decade has passed.
c. Such kind of feature efficiency is less reliable as it has to rely on human input.

15.2 Proposed System

The proposed self-operating light dimming (hereby referred as S.O.L.D.) is completely self-dependent and works without human dependency.

This system relies on the input data from:

i. **BNO055 9-DOF**: This high-precision 3-axis orientation sensor has incorporated an accelerometer, gyroscope, and magnetometer to output accurate and lives 3D position of vehicle [5–8]. This helps to know the current driving conditions, i.e., incline/decline or turning of the vehicle.
ii. **BME280**: This high-precision sensor from Bosch has inbuilt temperature, humidity, and altimeter sensing that is utilized for calculation of loss in visibility on the road due to fog, smog, rainfall, dust storms, and similar vision obstructing conditions on the road [9–12].
iii. **HB100**: This robust microwave radar can detect objects up to 16 m [13–16] and will be used alongside other sensors to determine the beaming (high or low) to choose.

iv. **ALS-PT19**: High reliability light detection sensor to detect amount of ambient light in the environment to decide ON/OFF or level of headlight brightness [17–20].

15.2.1 Functioning of the Proposed System

The working of the sensor system that is fitted in the proposed system is explained with the help of following line diagrams.

15.2.1.1 3D Orientation Sensor

The system is placed on the car floor. It detects car tilt, inclination, decline of vehicle and weight depression. If any of the value is more than the given the threshold value, a signal in terms of percentage increase is sent to the headlight lift motor for coming in the same position. Block diagram of the same is shown in Fig. 15.1.

15.2.1.2 Temperature Humidity and Altitude Sensor

Temperature humidity and altitude sensor along with the same also detect the parameters rain, fog, dust, smog, and storms for better visibility. If the visibility is below the threshold level, power correction is provided to get required light intensity through head lamp. The same is explained with the help of the diagram in Fig. 15.2.

15.2.1.3 Presence and Distance of the Object in Front

This sensor is placed in front of the vehicle, and it detects presence and distance of the object in front. If the proximity is closer than threshold, then a low beam signal will be given to the head light. The same is explained with the help of process flow diagram in Fig. 15.3.

15.3 Advantages of Proposed Model

Following are the advantages of the proposed model:

i. **Cost effective**: The designed beam model is cost-effective model and can be fitted to any vehicle.
ii. **New high-precision technology**: New and high-precision sensors from well-known manufacturers ensure a reliable long-lasting operation.

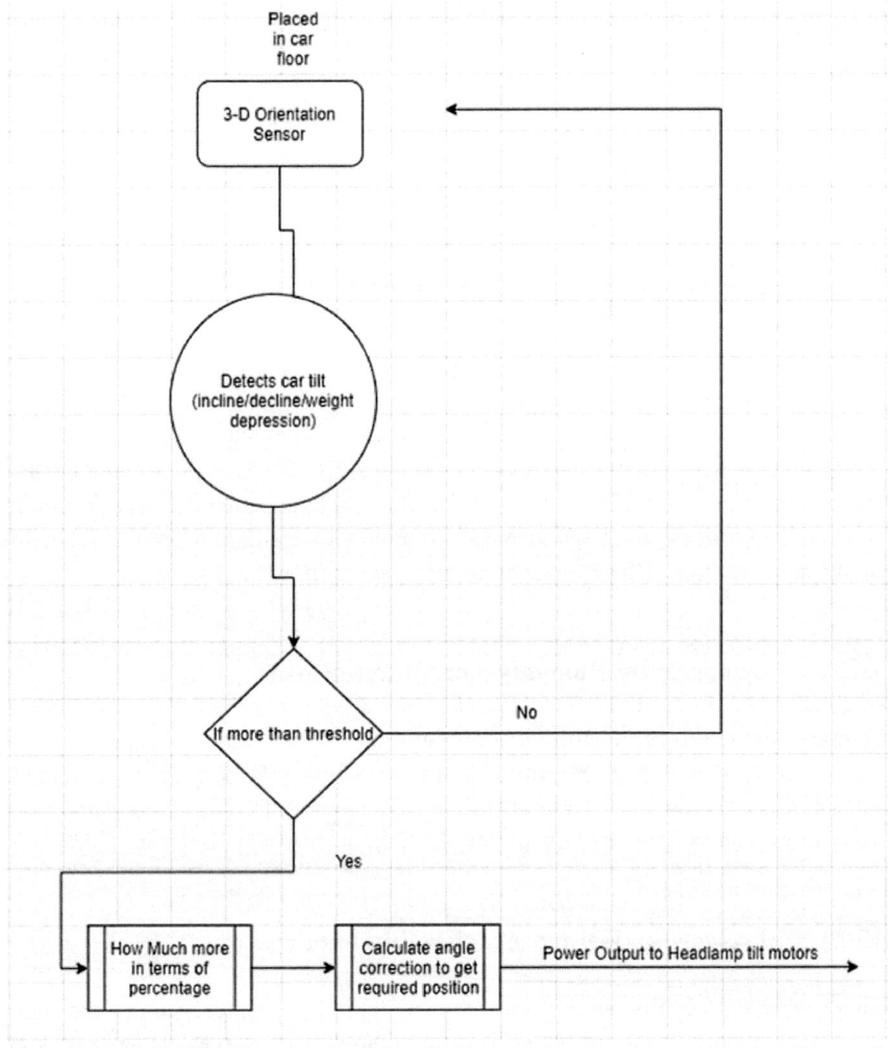

Fig. 15.1 Three-dimensional orientation sensor process flowchart

iii. **Efficiency**: With the fusion of modern high-efficiency sensors, the proposed model is more efficient than the pre-existing model in terms of precise detection of objects and necessary action with respect to obtained input parameters.

iv. **Control**: The proposed system is better in handling the lighting than the pre-existing system **sensing.** With more and accurate data from the sensors involved, the proposed system ensures that the lighting is always in the best position.

v. **Robust**: The proposed system is robust and functions as expected even in worst conditions.

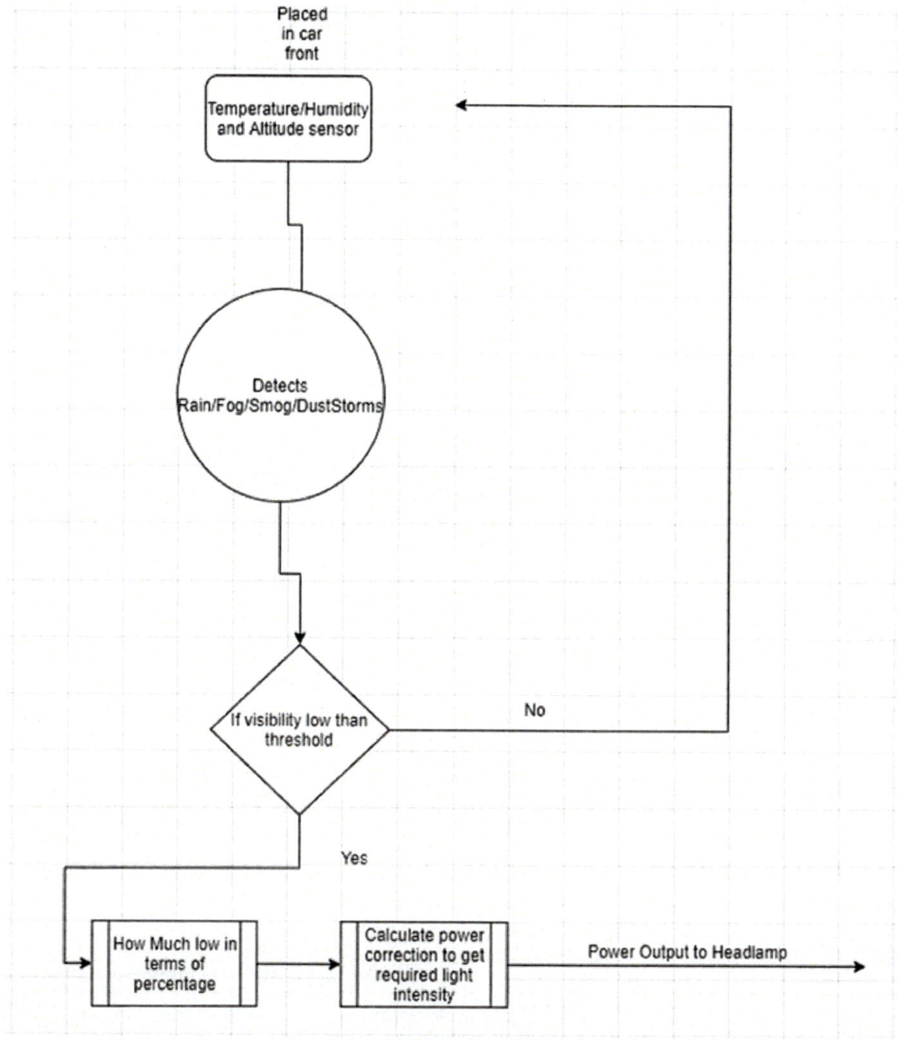

Fig. 15.2 Temperature humidity and altitude sensor process flow diagram

vi. **Fail proof**: The proposed system utilizes well-tested sensors from recognized brands and is expected to perform under extreme conditions with negligible failure rate.

vii. **All-in-one**: The proposed system not only has the ability of all other automated systems combined but more with better functioning and ease.

15.4 Comparison Analysis

A comparative analysis of existing and proposed system is shown in Table 15.1.

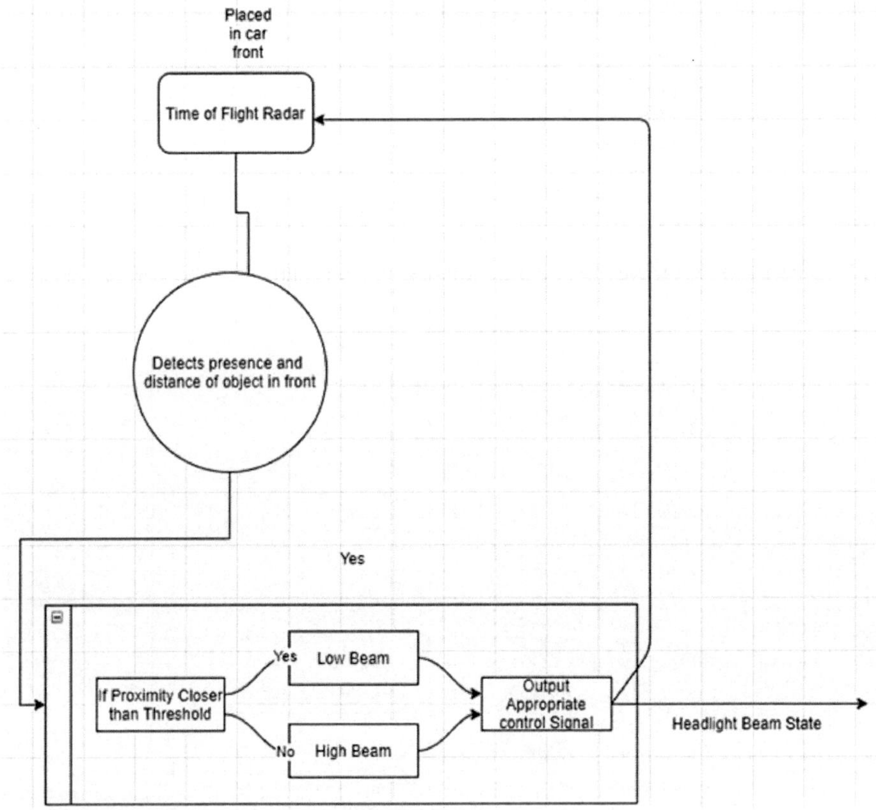

Fig. 15.3 Radar sensor process flow diagram

Table 15.1 A comparative chart for existing and proposed system

Feature	Existing system	Proposed system
Automatic headlight ON/OFF switching	Yes	Yes
[a] Light dimming to get required illumination	No	Yes
[a] Automatic headlamp adjustment	No	Yes
Light follow turns	Yes	Yes
[a] Incline/decline adjustment	No	Yes
Reliable operation during rain/fog	No	Yes
Best price/features	No	Yes
Automatic beaming control (high/low)	Yes	Yes
[a] Dynamic beaming (according with speed/road conditions)	No	Yes

[a] Separate system for only this particular system available. Not part of deployed automatic light control systems as a set

Table 15.2 Cost analysis of the designed system

Part/service	Price (in rupees)
ALS-PT19	285
HB-100	1526
BME280	340
BNO055	1000
Expected cost (expelling taxes and manufacturing cost)	3151

15.5 Cost Analysis

Installing such system will cost around Rs. 4000/ in the existing vehicles. The analysis related to cost is shown in Table 15.2.

15.6 Findings

In this, we find that as the technology gets advanced in the case of vehicles that work on the head light, most researchers are working on the safety of the driver or on the sight and power of the headlight. The problem of high beam is also solved with this kind of arrangement.

The proposed system worked well under the given constraints.

15.7 Conclusion

Following conclusion can be drawn from the study:

1. The problem of glare is taken care of by the proposed system
2. Proposed automatic headlight system is the new and latest technology
3. Audino sensor has to be provided with codings as per the external environment.
4. The system actuates and senses the light as at the daytime lights get automatically switched OFF and at night or at no light time headlights automatically gets switched ON. In this, we save our energy and cost of headlight.

References

1. R. Navaneethakrishnan, M. Santhanalakshmi, R. Marimuthu, A. Kumaresan, M. Alagumeenaakshi, ADAS Headlamp for improved visibility. Int. J. Pure Appl. Math. **119**(12), 12541–12548 (2018)

2. V. Anupama, D. Pankaj, G. Anand, K.P. Soman, Automatic anti-glare system for night time driving using liquid crystal screens. Int. J. Res. Eng. Technol. **03**(01), 202–205 (2014)
3. R. Dhawle, R. Anvekar, S. Kulkarni, S. Durg, Automatic headlight beam intensity switcher. Int. J. Res. Eng. Technol. **04**(05), 482–448 (2015)
4. R. Muralikrishnan, Automatic headlight dimmer a prototype for vehicles. Int. J. Res. Eng. Technol. **03**(02), 85–90 (2014)
5. K. Lakshmi, R. Nevetha, S.N. Ilakkiya, R. Ganesan, Automatic vehicle headlight management system to prevent accidents due to headlight glare. Int. J. Innov. Technol. Expl. Eng. **8**(9), 757–760 (2019)
6. K. Ashiq Ahamed, A.S. Ranganathan, Automatic dimming of headlights using vehicle speed. J. Chem. Pharm. Sci. (6), 4–7 (2015)
7. J.D. Bullough, J. Van Dorlofske, *Vehicle Forward Lighting: Optimizing for Visibility and Comfort. A Transportation Lighting Alliance Scoping Study* (Lighting Research Centre, Rensselaer Polytechnic Institute, Troy, NY, 2004), pp. 1–33
8. S.K. Okrah, E.A. Williams, F. Kumassah, Design and implementation of automatic light dimmer using LDR sensor. Int. J. Emerg. Technol. Innov. Eng. **2**(4) (2016)
9. F. Yakuphanoglu, Study and working performance of dimming methods of single and multichip power LED's. Int. J. Photoenergy **2012** (2012)
10. K.M. Folta, L.L. Koss, R. McMorrow, H.-H. Kim, J.D. Kenitz, R. Wheeler, J.C. Sager, Design and fabrication of adjustable red-green-blue LED light arrays for plant research. BMC Plant Biol. (2005)
11. K. Lakshmi, R. Nevetha, S.N. Ilakkiya, R. Ganesan, Automatic Vehicle Headlight Management System to prevent accidents due to headlights glare. Int. J. Innov. Technol. Expl. Eng. (IJITEE) **8**(9) (2019)
12. A. Varma, S. Kumar, R. Sai Varma, M. Sukumar, P Rajesh, Intelligent head light controller for vehicles. Int. J. Curr. Eng. Sci. Res. **5**(2) (2018)
13. A. Sutagundar, B. Patil, K.S. Srinidhi, Y. Yajaman, *Automated Headlight System Using Embedded Computing System* (BEC, Bagalkot, Karnataka/SJBIT, Bengaluru, Karnataka, 2018)
14. S.M. Gend, S.A. Jadhav, P.R. Sarode, A. Sharma, Adaptive driver assistance using automatic headlights. Int. J. Sci. Eng. Res. **8**(11) (2017)
15. D.K. Rath, Arduino based: smart light control system. Bhubaneswar Int. J. Eng. Res. Gen. Sci. **4**(2) (2016)
16. R. Muralikrishnan, Automatic headlight dimmer a prototype for vehicles. Int. J. Res. Eng. Technol.
17. J.J. Fazzalaro, Limitations on Headlight brightness, OLD research report. Br. J. Ophthalmol **87**(1), 113–117 (2003)
18. S.K. Choudhary, Electronic head lamp glare management system for automobile applications. Int. J. Res. Advent Technol. **2**(5), 402–416 (2014)
19. D. Roy Choudhary, *Linear Integrated Circuits*, 4th edn. (New Age International Publishers), pp. 311–315
20. C. Guttman, High Intensity Headlights could cause road accidents by dazzling oncoming drivers. Eurotimes (2003)

Chapter 16
Comparative Analysis of High-Gain Transformerless DC–DC Converter for DC Mircogrids

K. V. Suresh, K. U. Vinayaka, and N. V. Jyothi

Abstract Energy problems are increasing every day because of depletion of conventional fossil fuels. To overcome this problem, the microgrids play an important role. DC microgrids are very essential because of existence of photovoltaic cell, battery, and fuel cell. The DC microgrid will be more effective by the proper utilization of power electronic converters. In this paper, the characteristics of several DC–DC converters are analyzed with respect to parameters like duty cycle, voltage gain, and switching stress. The simulation of boost derived converters and non-isolated converters is carried out in PSIM.

Keywords DC micro grids · Boost derived converters · Non isolated converters · Voltage gain · Switching stress

16.1 Introduction

In India, energy conservation is very necessary because the sources that produce electricity are decreasing faster than generating. The growth of country depends on industrial and agricultural sector where they are major energy-consuming areas. In addition to the consumption, the major energy is consumed in terms of transmission and distribution losses such as energy dissipation in conductors, theft of energy, and meter tampering. To overcome these losses, the demand-side management is required which enables the smart building by suggesting on-site generation such as microgrids. Nowadays, the DC loads are increasing such as LED lights, laptop batteries, and cell phone batteries. Hence, DC microgrids are dominating because of existence of photovoltaic panel and energy storage devices such as battery and fuel cell.

K. V. Suresh
Siddaganga Polytechnic, Tumakuru, India

K. U. Vinayaka (✉) · N. V. Jyothi
Siddaganga Institute of Technology, Tumakuru, India
e-mail: vinayaka@sit.ac.in

A. Khosla and M. Aggarwal (eds.), *Smart Structures in Energy Infrastructure*,
Studies in Infrastructure and Control,
https://doi.org/10.1007/978-981-16-4744-4_16

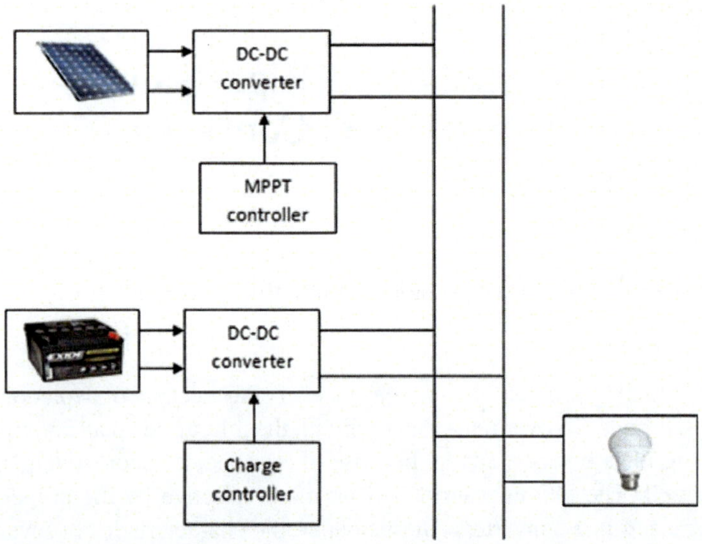

Fig. 16.1 Block diagram of DC microgrid

The DC microgrids are having following advantages:

1. It reduces energy conversion losses and improves the energy efficiency
2. Synchronization problem can be eliminated
3. Increases the reliability for emergency loads in hospitals and financial institutions
4. Reduces the construction cost of new power plants by reducing peak demand.

16.2 Block Diagram

The functional block diagram of DC microgrid is as shown in Fig. 16.1. The block diagram consists of solar panel, battery unit, controllers, DC bus, and DC loads. In DC microgrid, generally two sources are used: one may be renewable energy source and other one must be energy storage device. In the block diagram, the PV panel and battery unit are taken as input source. A power electronic converter such as DC-DC converter is used as an interface between the source and DC bus. Choice of the converter is based on input source. Along with source and converters, the MPPT controller is used at renewable energy side and charge controller is used at energy storage source side.

16.3 Converters in DC Microgrid

The different converters are used for DC microgrids are as follows.

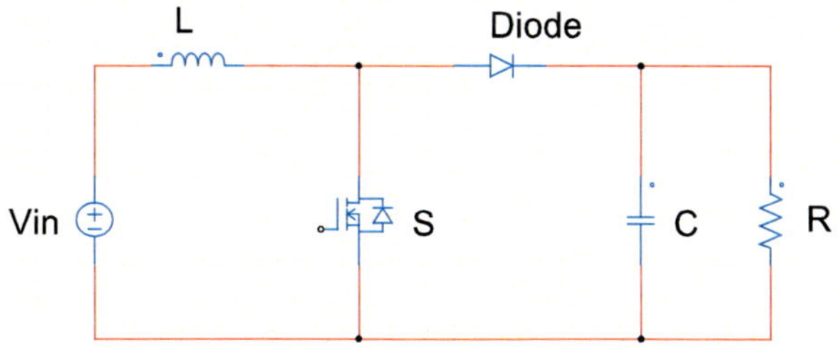

Fig. 16.2 Circuit of conventional boost converter

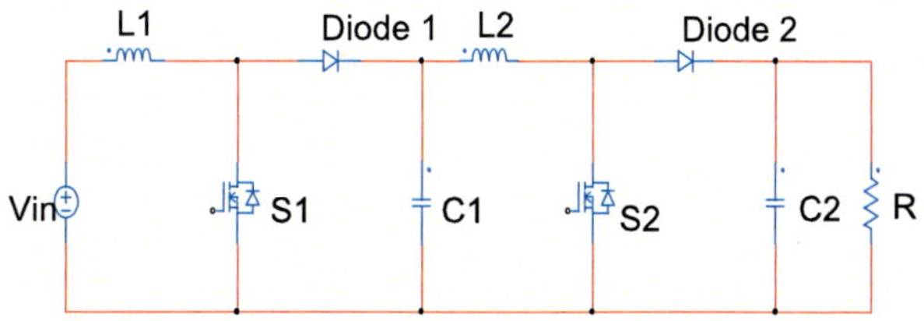

Fig. 16.3 Circuit of CBC

16.3.1 Conventional Boost Converter

The circuit representation of conventional boost converter is as displayed in Fig. 16.2, and the circuit comprises of an input DC voltage (V_{in}), power electronic switch (S), and circuit elements like inductor (L) and capacitor (C). The converter operation is described in two modes. First when S is closed then diode is reverse biased. When S is open the inductor current forces diode [1].

16.3.2 Cascaded Boost Converter (CBC)

The circuit diagram of CBC is as shown in Fig. 16.3. It consists of input voltage (V_{in}), switches (S1 & S2), diode 1 and diode 2, inductors (L1 & L2), and capacitors (C1 & C2). This converter comprises of two boost converters in series. The converter is operated in two modes. When both switches are ON, then the diodes operate in reverse biased. With cascade connection, the desired gain in the voltage can be achieved easily [2].

16.3.3 Dual Boost Converter (DBC)

The circuit diagram of DBC is as shown in Fig. 16.4. It consists of input voltage (V_{in}), switches (S1 & S2), diode 1 and diode 2, inductors (L1 & L2), and capacitors (C1 & C2). Two units of boost converters are connected in parallel to realize a CBC. The converter is operated in two modes. When both switches are ON, then the diodes

Fig. 16.4 Circuit of DBC

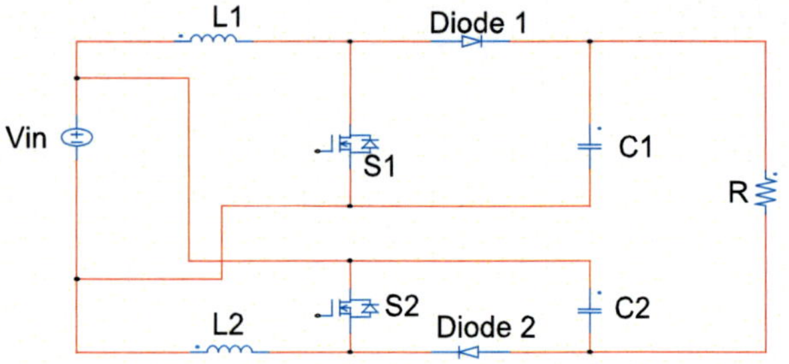

Fig. 16.5 Circuit of BVMC

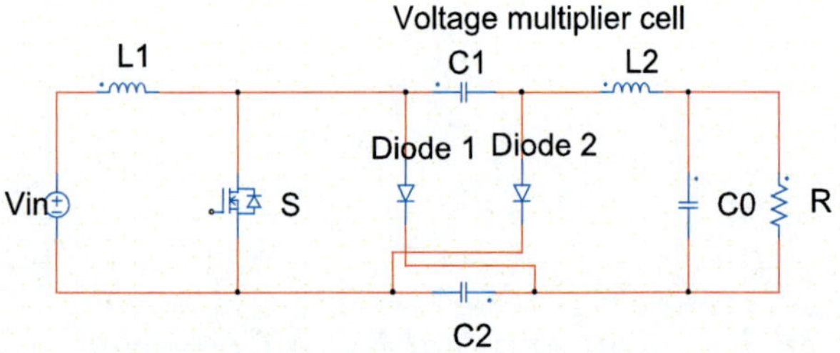

are reverse biased. With parallel connection, the voltage and current stress can be reduced [2].

16.3.4 Boost with Voltage Multiplier Cell (BVMC)

The circuit diagram of BVMC is as shown in Fig. 16.5. It consists of voltage (V_{in}), switch (S), inductors (L1 & L2), and capacitor (C). Along with the above components, it also consists of voltage multiplier cell (diode 1, diode 2, capacitors C1 & C2). With this converter, the switching stress can be reduced and voltage gain can be increased. This circuit utilizes only one switch. When S is closed, the diodes are reverse biased. When S is opened, the diodes are forward biased [2].

16.3.5 Non-Isolated Boost Converter 1 (NIBC1)

The above circuit is a modified boost converter with coupled inductors (L1 & L2) having same number of turns. It consists of three switched (S1, S2, & S3) and capacitor (C). The circuit can be operated in both step-up and step-down mode by interchanging input voltage and load resistance positions. The switches S1 and S2 are ON for certain duration and switch S3 is ON for remaining duration over a switching period [3] (Fig. 16.6).

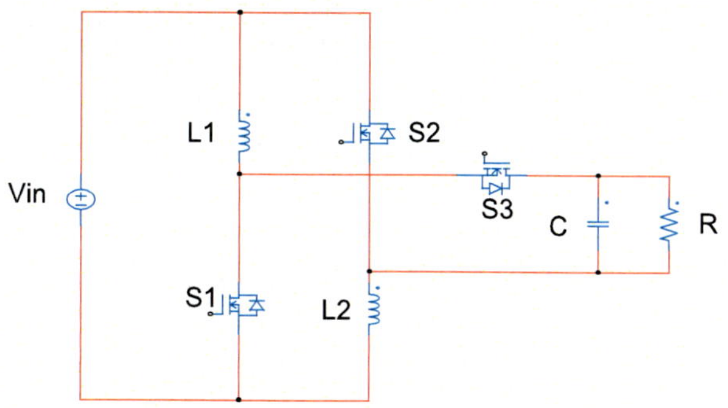

Fig. 16.6 Circuit of NIBC1

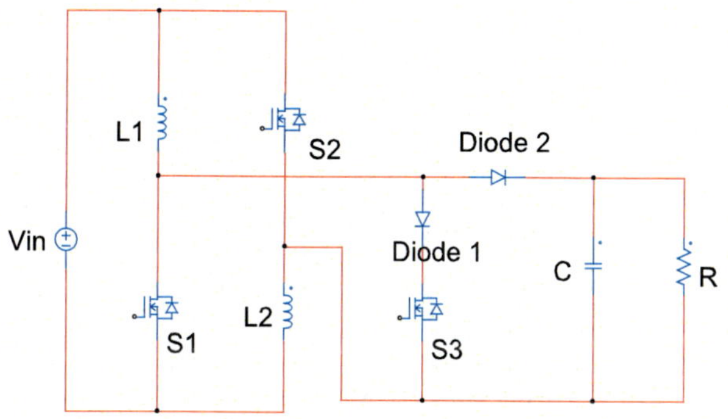

Fig. 16.7 Circuit of NIBC2

16.3.6 Non-isolated Boost Converter 2 (NIBC2)

The circuit diagram of NIBC2 is as shown in Fig. 16.7. The circuit is having coupled inductors (L1 & L2) having same number of turns, three switched (S1, S2 & S3), diode 1, diode 2, and capacitor (*C*). The circuit is similar to non-isolated boost converter with improved voltage gain. The circuit is operated in three modes such as

- When switches S1 and S2 are ON, then both diodes operate in reverse bias. The duty cycle for this duration is taken as D1
- When switch S3 is turned ON, then the diode 1 is forward biased and diode 2 is reverse biased. The duty cycle for this duration is taken as D2
- When all the switches are OFF, then diode 1 is reverse biased and diode 2 is forward biased. The interval of this mode can be taken as (1-D1-D2) [4].

16.4 Comparison of Converters

The simulation work of different converters is carried out in PSIM. The design is considered by taking input voltage as 20 V and output voltage as 200 V, 100 W power. The switching frequency for all the converters is taken as 50 kHz.

The simulation results of boost derived converters are listed in Table 16.1.

Table 16.1 Comparison of converters for different parameters

S. No.	Parameter	Boost converter	Cascaded boost converter	Dual boost converter	Boost converter with voltage multiplier cell	Non-isolated converter 1	Non-isolated convener 2
1	Input voltage (in volts)	20	20	20	20	20	20
2	Output voltage (in volts)	200	200	200	200	200	200
3	Switching frequency (in kHz)	50	50	50	50	50	50
4	Duty cycle	0.9	0.68	0.8181	0.8181	0.8181	0.5
5	Voltage gain factor	$\frac{1}{(1-D)}$	$\frac{1}{(1-D)^2}$	$\frac{(1+D)}{(1-D)}$	$(1+D)$	$(1+D)$	$\frac{1}{(1-D_1-D_2)}$
6	Inductors (in μH)	720	720, 720	720, 720	720	65, 650	360, 360
7	Capacitors (in μF)	90	90, 90	90, 90	10, 10, 90	82	50
8	Number of switches	1	2	2	1	3	3
9	Voltage across switches	200	63, 200	110, 110	110	110, 110, 220	105, 105, 190
10	Current through switches	5	5, 2	265, 265	5	2.8, 2.8, 2.8	3.2, 12, 3.6
11	Polarity	Same	Same	Same	Reverse	Same	Same

With the comparison, the high voltage gain is achieved by NIBC2. The low switching stress can be achieved by BVMC, and this converter can be implemented with only one switch.

Figure 16.8 represents variation of duty cycle for different converters. The boost converter requires high duty cycle. The cascaded boost converter requires lesser duty cycle. The duty cycle of non-isolated converter 2 is very least. The remaining converters are having same duty cycle.

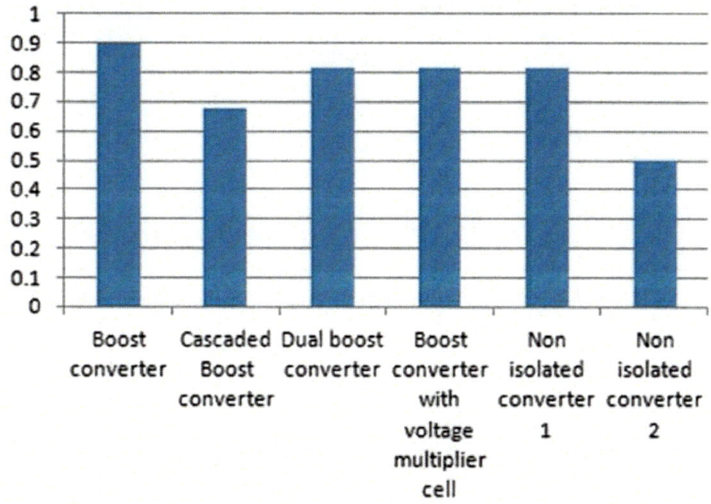

Fig. 16.8 Duty cycle requirement of different converters

16.5 Simulation Results

The performance analyses of different converters are carried out by simulation using PSIM as per the design. The relevant waveforms of input and output voltage pertaining to a boost converter are depicted in Fig. 16.9, and switching stress waveforms are shown in Fig. 16.10. The potential difference across the switch is 200 V which is equal to output voltage, and current through switch is 5 A.

The input voltage and output voltage waveforms of CBC are shown in Fig. 16.11, and switching stress waveforms are shown in Figs. 16.12 and 16.13. The voltages across the switches are 63 and 200 V, and current through switches 5 and 2 A.

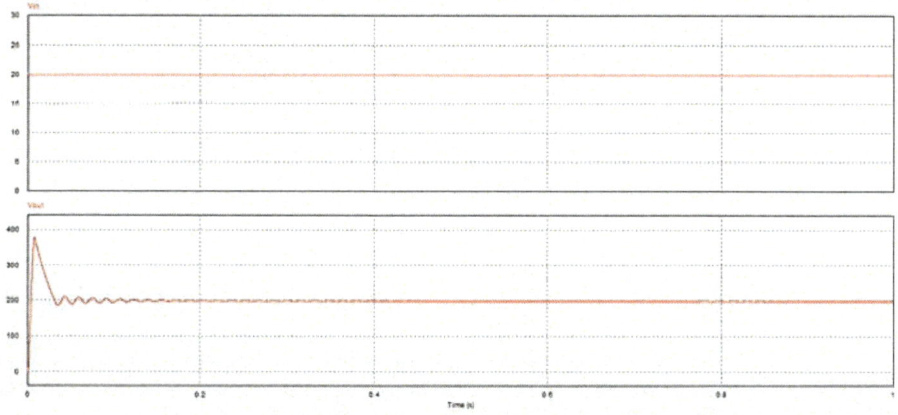

Fig. 16.9 Input voltage and output voltage waveforms of conventional boost converter

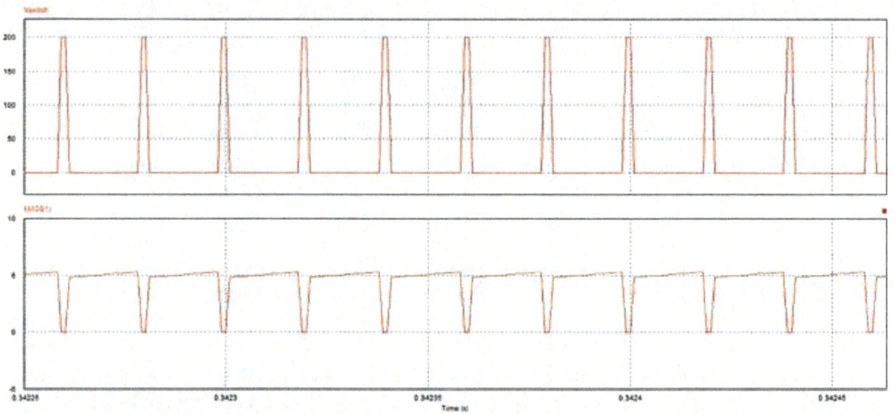

Fig. 16.10 Voltage across switch and current through switch for boost converter

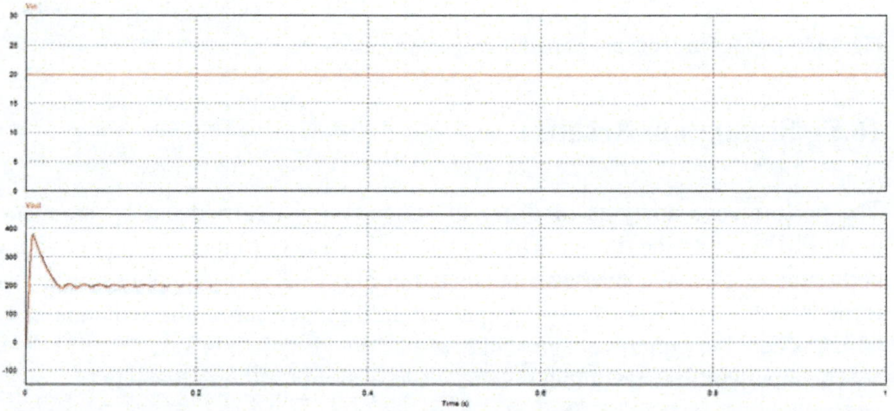

Fig. 16.11 Input voltage and output voltage waveforms of CBC

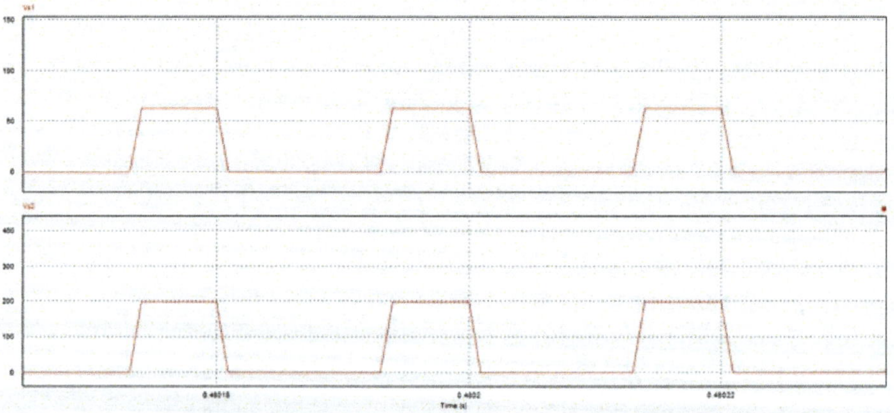

Fig. 16.12 Voltages across S_1 and S_2 of CBC

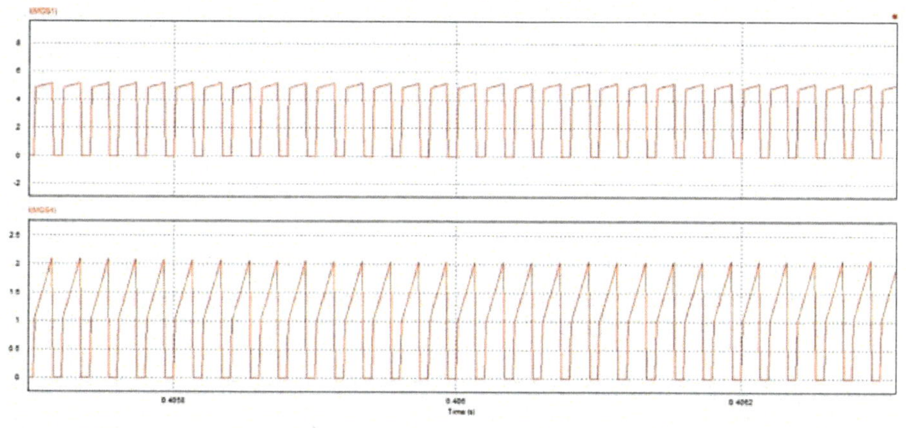

Fig. 16.13 Current through S_1 and S_2 of CBC

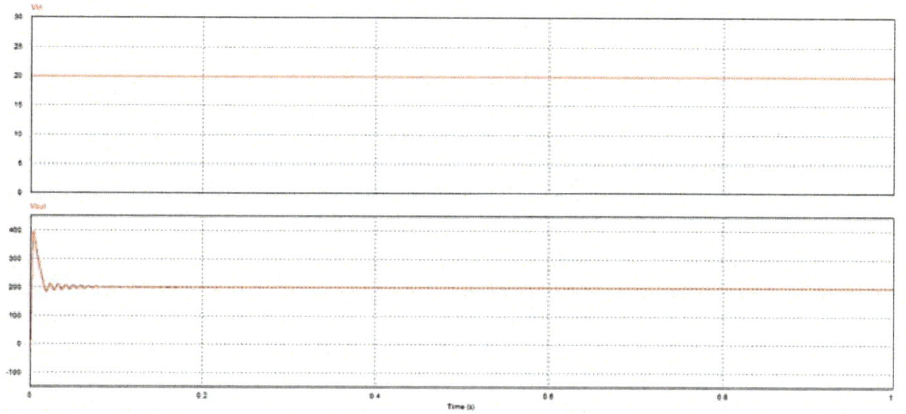

Fig. 16.14 Input voltage and output voltage waveforms of DBC

The input and output voltage waveforms of DBC are shown in Fig. 16.14, and switching stress waveforms are shown in Figs. 16.15 and 16.16. The voltages across the switches are 110 V in each and current through switches 2.65 A in each.

Waveforms of input and output voltage of boost converter with voltage multiplier cell is represented in Fig. 16.17, and switching stress waveforms are shown in Fig. 16.18. The voltage across the switches is 110 V, and current through switch is 5 A.

The input and output voltage waveforms of non-isolated converter are shown in Fig. 16.19, and switching stress waveforms are shown in Figs. 16.20 and 16.21. The voltage across the switches is 110 and 220 V, and current through switch is 2.8 A.

The input and output voltage waveforms of non-isolated converter are shown in Fig. 16.22, and switching stress waveforms are shown in Figs. 16.23 and 16.24. The voltage across the switches is 105 and 190 V, and current through switches are 3.2 and 3.6 A.

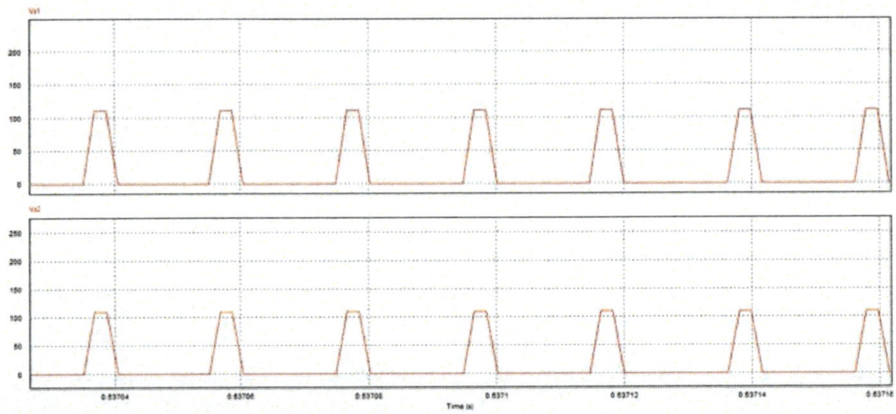

Fig. 16.15 Voltages across S_1 and S_2 of DBC

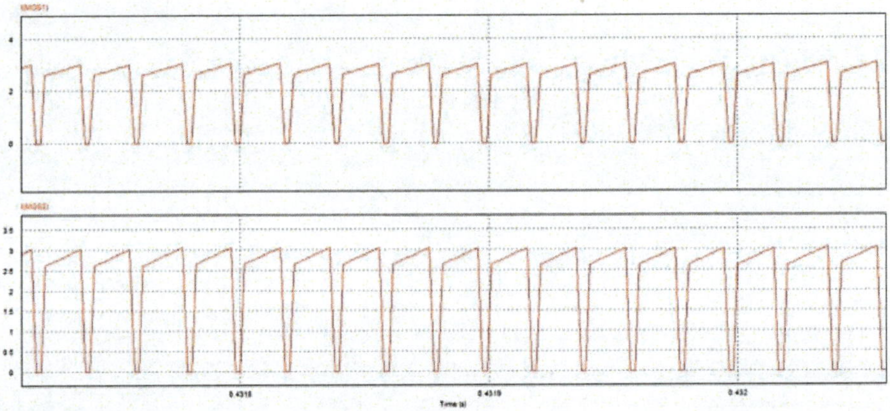

Fig. 16.16 Current through S_1 and S_2 of DBC

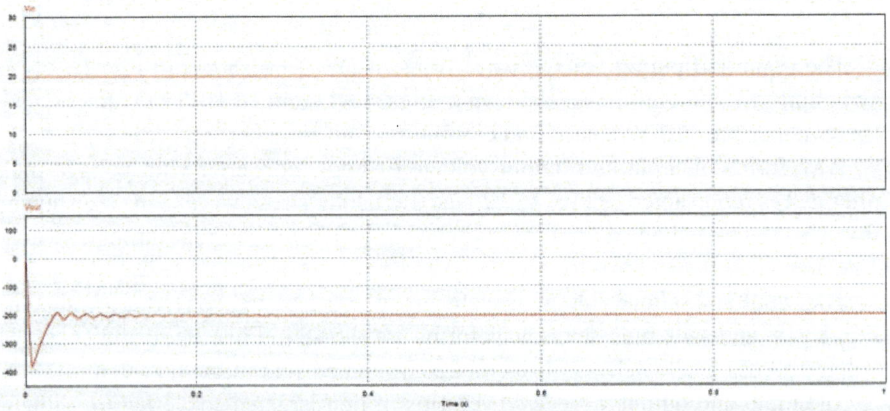

Fig. 16.17 Input voltage and output voltage waveforms of boost converter with voltage multiplier cell

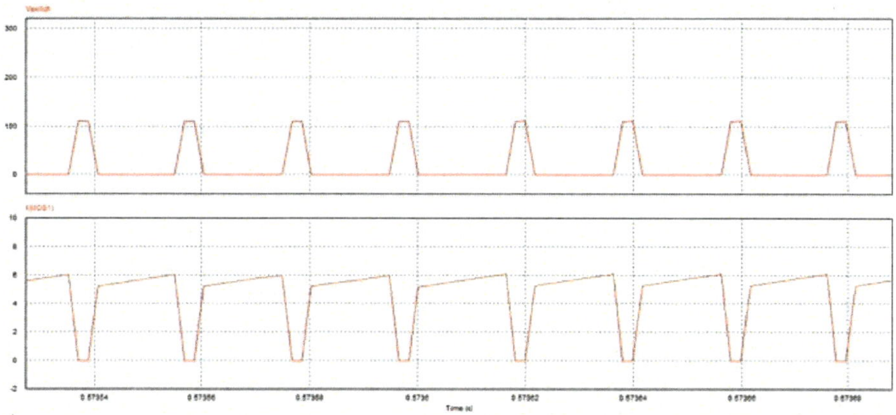

Fig. 16.18 Voltage across switch and current through switch for BVMC

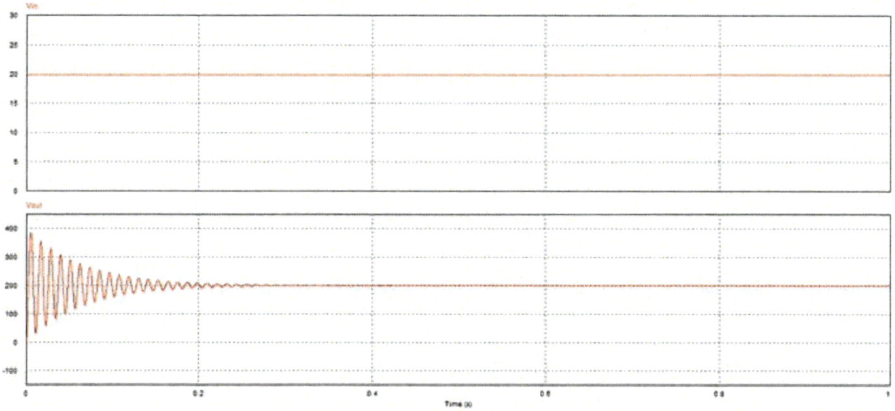

Fig. 16.19 Input voltage and output voltage waveforms of non-isolated converter 1

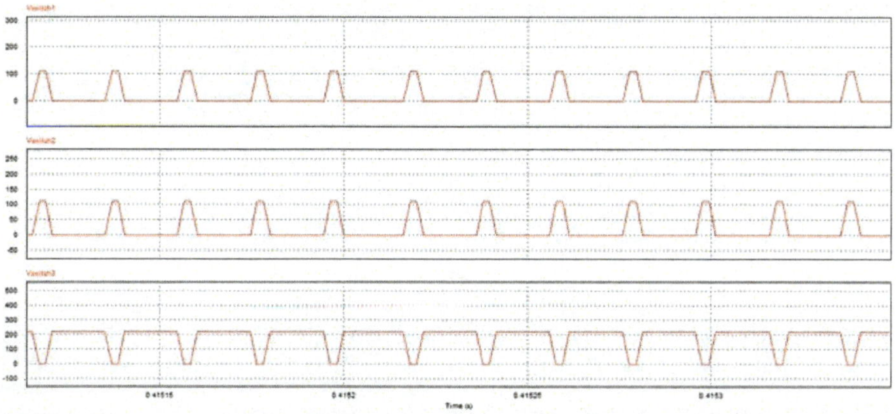

Fig. 16.20 Voltage across switch of non-isolated converter 1

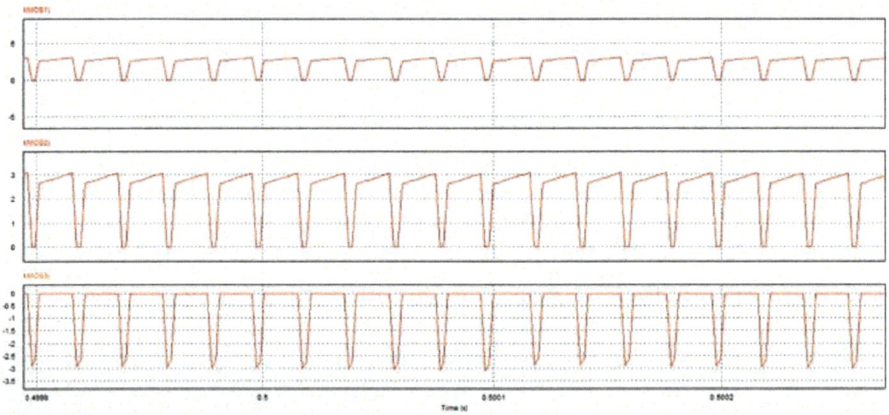

Fig. 16.21 Current through switches in non-isolated converter 1

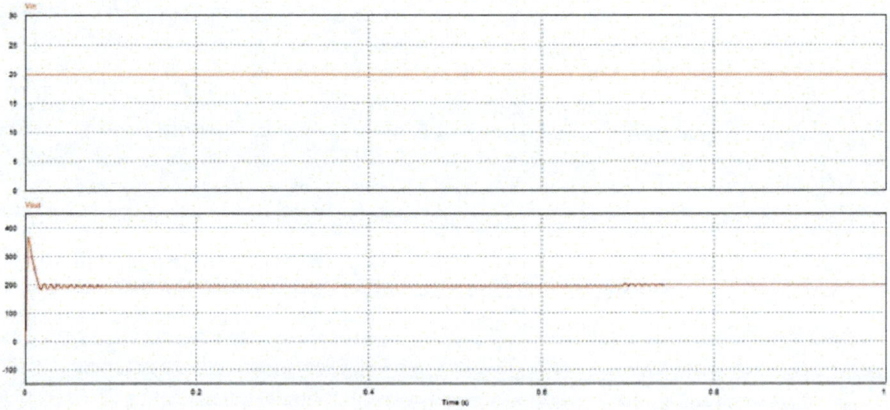

Fig. 16.22 Input and output voltage waveforms of non-isolated converter 2

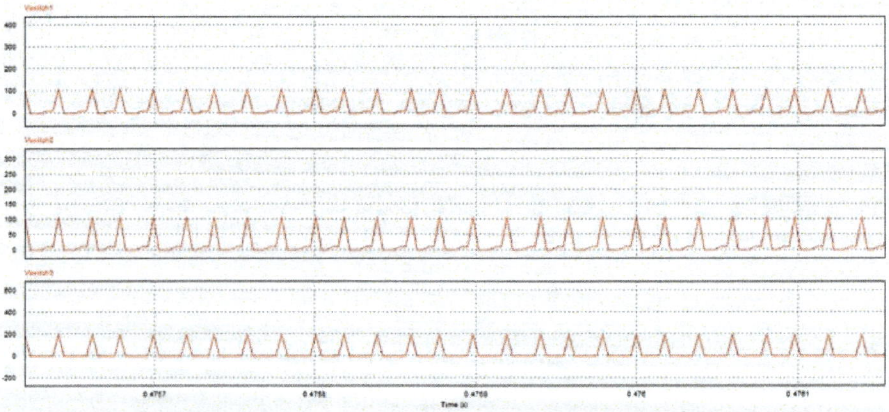

Fig. 16.23 Voltage across switch of non-isolated converter 2

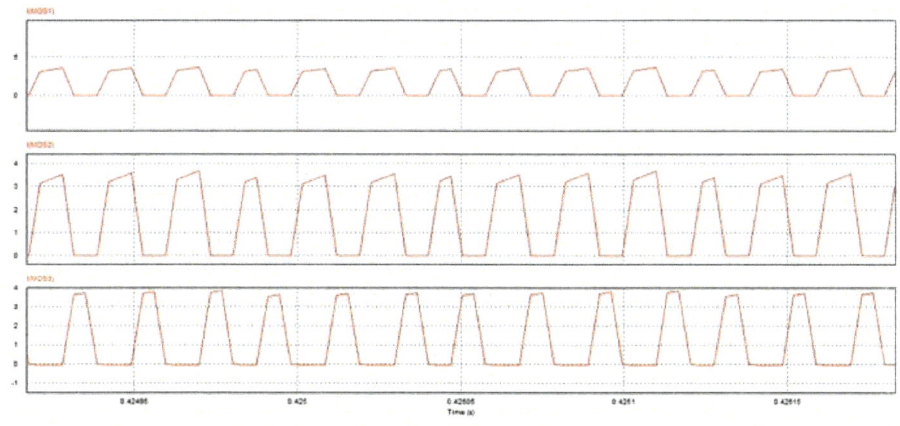

Fig. 16.24 Current through switches in non-isolated converter 2

16.6 Conclusion

In this work, the operation of different converters that analyzed the various parameters of high-gain DC-DC converters for DC microgrid is carried out. Each every converter has its advantages:

- The high voltage gain is achieved by non-isolated converter 2
- The current stress can be reduced by non-isolated converter 1
- The BVMC cell requires only 1 switch
- The voltage stress is decreased by cascaded boost converter
- The conventional boost converter requires less number of passive elements.

References

1. D.W. Hart, *Power Electronics*, 1st edn. (McGraw-Hill Education). ISBN-10: 0073380679
2. J. Divya Navamani, K. Vijaya kumar, A. Lavanya, A. Jason Mano Raj, Non-isolated high gain DC–DC converter for smart grid—a review, in *NCMTA 2018*
3. L.-S. Yang, T.-J. Liang, Analysis and implementation of a novel bidirectional DC–DC converter. IEEE Trans. Ind. Electron. (2012)
4. Lakshmi, S. Hemamalini, Nonisolated high gain DC–DC converter for DC microgrids. IEEE Trans. Ind. Electron. (2018)

Chapter 17
Energy Harvesting with Photovoltaic Arrays: Assessment of Reliability with Alternative Configurations for Power Delivery

Sandhya Prajapati, Yuvraj Praveen Soni, and E. Fernandez

Abstract For evaluation of operational suitability of a system, reliability is considered as an important parameter for the analysis. Evaluation of reliability of photovoltaic arrays (PVA) operation is necessary for critical power applications in which a failure of one or more modules of the array may occur. This study discusses the issues of solar photovoltaic (SPV) stand-alone reliability behavior when PV array is arranged in different series–parallel modular configurations This paper focuses on the reliability assessment of different PVA containing ($m \times n$) single crystalline silicon solar cell modules which constitute the array, where m is the no. of modules in series and n is the no. of parallel strings involved. Probability theory has been implemented as to reduce complexity involved in PVA network to evaluate the reliability index used for the assessment.

Keywords Solar energy harvesting · Reliability · Solar PV array configuration · Probabilistic theory · Reliability index

17.1 Introduction

Solar photovoltaic (SPV) transforms the sun energy to electric energy which is a pollution-free, quite, and clean source of power. Photovoltaic (PV) modules are considered as extremely well-founded source of electrical power. To supply the power in unelectrified isolated or remote areas, PV power is recognized as practical alternatives. However, the practical scenarios show the failure or degradation of PV modules can occur in number of ways. According to literature [1, 2] and theories, the PV modules are expected to have life span of around 20 years; however in actual event, it reduces to 8–10 years due to packaging technology used for module and other related issues leading to damage of PV modules.

There are two possible ways which can cause the failure of solar PVA. In the first case, regular depreciation due to wear and tear and other natural causes related

S. Prajapati · Y. P. Soni (✉) · E. Fernandez
Indian Institute of Technology Roorkee, Roorkee, India

A. Khosla and M. Aggarwal (eds.), *Smart Structures in Energy Infrastructure*,
Studies in Infrastructure and Control,
https://doi.org/10.1007/978-981-16-4744-4_17

to ambient atmospheric conditions [3] cause regular degradation of PVA over a certain span. Various causes of wear and tear involve mechanism external to the cell itself such as encapsulant browning, solder joints, interconnection, and delamination problems. Various literatures are available which show the studies in the field that relate to PV module degradation [4–13]. This studies highlight the main conclusion showing that the solar modules do not fail in a disastrous way but do experience a degradation in efficiency over a time span. The leading issues identified are (a) packaging material degradation, (b) loss of adhesion of encapsulants, (c) degradation of module/cell interconnection, (d) degradation causes due to moisture intrusion, and (e) degradation of semiconductor device. In addition to these, the causes also involve detrimental ambient weather condition involving humidity, temperature, and radiation.

The second case which leads to the failure of SPV array is the case when any of the cell in the series modular string becomes open circuited or short circuited. In the short-circuited case, the modular will still deliver power, albeit at a lower level, if the configuration does not permit a total bypass of the load. However, in the case of an open circuit, the entire module becomes ineffective and the delivered power of the module series string is zero. Affected modules turn out to be non-operative which leads to a reduction in total power output of entire array. Solar PVA open-circuit conditions can occur in number of ways. One such related example is obstruction of solar rays before falling PVA. Potential causes include shadows of tree and cloud, poles, buildings, and various other objects, falling of paint, dirt, dirt layer (after a dust storm), bird excreta, and the cracking of protective glass covering, etc. Reliability of solar PV modules is therefore of concern.

17.2 Reliability Under Probabilistic Conditions

Under the nominal situations, the solar cell efficiency degrades in an exponential manner. Under the field condition, the failure of the solar module may occur anytime due to the environmental stresses or may survive functioning till the end of its predicted lifespan [14].

Let us consider a solar PV array of dimensions $m \times n$, where m is the no. of modules in series and n is the no. of parallel strings involved. Let Xij represent any module in the associated array matrix. The probability of the module Xij to function accurately for time t is represented by $P(X.t)$. The probability of failure of cell within time t is represented by $Q(X.t)$.

The cell will be surely at either of the condition which sums up the malfunction and accurate operation probability to be equal to 1, i.e., $P(X.t) + Q(X.t) = 1$. The measured probability of Xij solar module to operate properly till time t represents the reliability of Xij module.

The above relation can be expressed mathematically as $R(t) = P(X.t) > t = 1 - Q(t)$. $Q(t)$ represents the distribution function of the cell lifetime (Fig. 17.1).

The reliability index $R_k(t)$ is given as [15]:

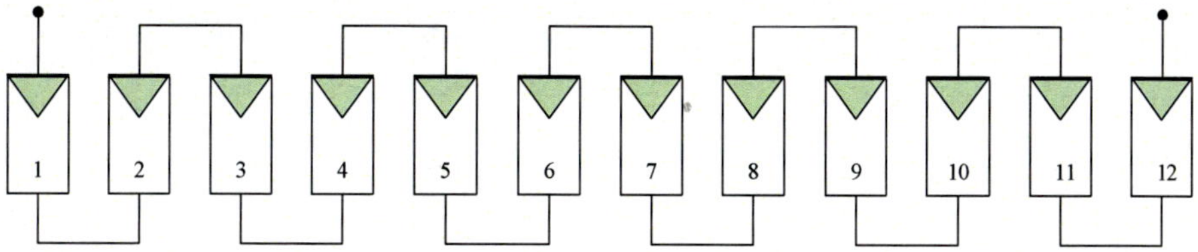

Fig. 17.1 A series string of a solar PV modules

$$R_k(t) = P(X_k > t) \tag{17.1}$$

$$R_k(t) = P((X_1 > t) \cap (X_2 > t) \ldots \cap (X_m > t) \tag{17.2}$$

$$R_k(t) = P(X_1 > t)P(X_2 > t)P(X_3 > t) \ldots P(X_m > t) \tag{17.3}$$

$$R_k(t) = R_1(t)R_2(t) \ldots R_m(t) \tag{17.4}$$

$$R_s(t) = \prod_i^m R_i(t) \tag{17.5}$$

For a parallel combination, $m = 1$ while $n =$ some finite value (Fig. 17.2).

The failure of all n branches of the system will lead to failure of parallel system. If each branch consists of a single module then, probability for failure of parallel system with n modules can be given as

$$= (1 - P(X_1 > t))(1 - P(X_2 > t))(1 - P(X_3 > t)) \ldots (1 - P(X_n > t)) \tag{17.6}$$

$$= \prod_i^n Q_i(t) \tag{17.7}$$

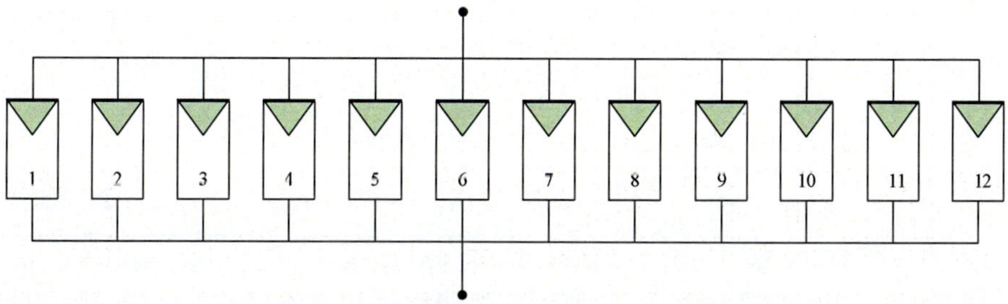

Fig. 17.2 A parallel set of solar PV modules

The reliability index $R_p(t)$ is given as [15]. Since $Q_i(t) = 1 - R_i(t)$, we have

$$R_p(i) = 1 - \prod_{i}^{n} (1 - R_i(i)) \quad (17.8)$$

A practical solar PV array involves $m > 1$ and $n > 1$. In such cases, it can be shown [14] that:

$$R_{sp}(i) = 1 - \prod_{i}^{n} (1 - R_{ij}(i)) \quad (17.9)$$

The object of the present paper is to understand how varying combinations of m and n can influence the reliability of the solar PV array.

17.3 Simulation Methodology

The simulation involves the reliability estimation of solar PV arrays with different combinations of m and n. Table 17.1 shows the various values of m and n that have been used to produce different combinations for the array.

Using the formulae for reliability estimation as given in Sect. 17.2, the following reliability values have been obtained as shown in Tables 17.2 and 17.3, respectively.

Table 17.1 Values of m and n used for array configuration

S. No.	m	n
1	1	1
2	2	2
3	3	3
4	4	4
5	5	5
6	8	8
7	10	10

Table 17.2 Reliability value of array for different combinations of m and n

Parallel n	Series m				
	3	4	5	8	10
1	0.729	0.656	0.594	0.43	0.348
2	0.927	0.881	0.832	0.675	0.576
3	0.98	0.959	0.931	0.815	0.724
4	0.994	0.986	0.972	0.895	0.82

Table 17.3 Reliability value of array for different combinations of n and m

Series m	Parallel n					
	2	3	4	5	8	10
1	0.99	0.999	0.999	0.9999	1	1
2	0.964	0.993	0.998	0.999	0.999	1

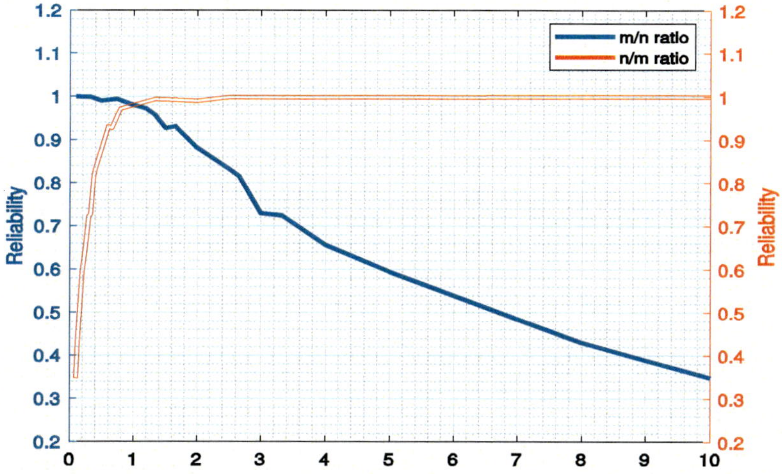

Fig. 17.3 Variation of reliability with **a** m/n ratio and **b** n/m ratio

It is also of interest to understand how the ratios m/n and n/m will influence the reliability. Figure 17.3 illustrates this:

17.4 Results and Discussion

The findings of Tables 17.1 and 17.2 indicate the following:

- As m increases, the reliability for a given value of n tends to fall. In other words, series string with greater no. of modules tend to be less reliable.
- Although the reliability falls with increase in m, yet a higher value of n will result in a decreasing value of m starting at higher levels of reliability. Figure 17.3 indicates that: Reliability falls as m/n ratio increases while reliability improves as n/m ratio increases.

Thus, we see that the ration of series to parallel modules and vice versa will have an impact in deciding the reliability level of the solar PV array. Thus, energy planners must plan appropriately to ensure the best level of reliability in energy harvesting of solar PV power.

17.5 Conclusion

The present paper is an attempt to understand by simulation how the combination of the no. of modules in series and in parallel can influence the reliability of the solar PV array. It was seen that reliability falls as *m*/*n* ratio increases while reliability improves as *n*/*m* ratio increases. Based on this study, the appropriate combination of *m* and *n* needs to be selected for best results in relation to a desired output power level.

References

1. S.E. Forman, Performance of experimental terrestrial photovoltaic modules. IEEE Trans. Reliab. **R-31**, 235–245 (1982)
2. J.H. Wohlgemuth, D.W. Cunningham, A.M. Nguyen, J. Miller, Long term reliability of PV module, in *20th European Photovoltaic Solar Energy Conference*, Barcelona, Spain (2005), pp. 1942–1946
3. M.A. Quintana, D.L. King, T.J. McMahon, R.C. Osterwald, Commonly observed degradation in field-aged photovoltaic modules, in *29th Photovoltaic Specialists Conference* (IEEE, Louisiana, 2002), pp. 1436–1439
4. S. Sakamoto, T. Oshiro, Dominant degradation of crystalline silicon photovoltaic modules manufactures in 1990, in *20th European Photovoltaic Solar Energy Conference* (Barcelona, 2005), pp. 2155–2158
5. I.J. Muirhead, B.K. Hawkins, An assessment of photovoltaic power in the Telstra network, in *Solar '95—Proceedings of the Annual Conference of the Australian and New Zealand* (Solar Energy Society, Hobart, Australia, 1995), pp. 493–500
6. A.M. Reis, N.T. Coleman, M.W. Marshall, P.A. Lehman, C.E. Chamberlain, Comparison of PV module performance before and after 11-years of field exposure, in *29th IEEE Photovoltaic Specialists Conference* (IEEE, New Orleans, Louisiana, USA, 2002), pp. 1432–1435
7. C.R. Osterwald, J.P. Benner, J. Pruett, A. Anderberg, S. Rummeland, L. Ottoson, Degradation in weathered crystalline-silicon PV modules apparently caused by UV radiation, in *3rd World Conference on Photovoltaic Energy Conversion*, Osaka, Japan (2003), pp. 2911–2915
8. S. Sakamoto, T. Oshiro, Field test results on the stability of crystalline silicon photovoltaic modules manufactured in the 1990, in *3rd World Conference on Photovoltaic Energy Conversion*, Osaka (2003), pp. 1888–1891
9. E.D. Dunlop, D. Halton, The performance of crystalline silicon photovoltaic solar modules after 22 years of continuous outdoor exposure. Prog. Photovoltaics Res. Appl. **14**, 53–64 (2006)
10. B. Marion, J. Adelstein, Long-term performance of the SERF PV systems, in *NCPV and Solar Program Review Meeting*, Colorado (2003), pp. 1–6
11. Z. Liu, M.L. Castillo, A. Youssef, J.G. Serdy, A. Watts, C. Schmid, S. Kurtz, I.M. Peters, T. Buonassisi, Quantitative analysis of degradation mechanisms in 30-year-old PV modules. Sol. Energy Mater. Sol. Cells **200**(1), 110019 (2019)
12. A. Omazic, G. Oreski, M. Halwachs, G.C. Eder, C. Hirschl, L. Neumaier, G. Pinter, M. Erceg, Relation between degradation of polymeric components in crystalline silicon PV module and climatic conditions: a literature review. Sol. Energy Mater. Sol. Cells **192**, 129–133 (2019)
13. M.A. Islam, M. Hasanuzzaman, N.A. Rahim, Investigation of the potential induced degradation of on-site aged polycrystalline PV modules operating in Malaysia. Measurement **119**, 283–294 (2018)
14. N.K. Gautam, N.D. Kaushika, Reliability evaluation of solar photovoltaic arrays. Sol. Energy **72**(2), 129–141 (2002)

15. R. Billinton, R.N. Allan, *Reliability Evaluation of Engineering Systems: Concepts and Techniques* (Plenum Press, Springer, New York, 1992)

Chapter 18
A Sensors-Based Solar-Powered Smart Irrigation System Using IoT

Sandeep Chopade, Swati Chopade, and Sushopti Gawade

Abstract With the availability of low cost, low power, and highly portable sensors for monitoring soil status, it becomes easier and convenient to assess the condition of soil or to identify dampness level. Information about the dampness level helps to irrigate the soil. In this work, we propose a sensor-based soil moisture content detection system using microcontroller. The system first presents sensing the moisture level of the soil. Next, the system presents the comparison between the sensed moisture level and the preset values stored in the microcontroller. Further, if the moisture level of the soil is less than threshold value, then the relay starts the solar-powered pump motor to irrigate the soil. Finally, GSM module is used to send or receive messages between microcontroller and the mobile owner. We evaluate the performance of the proposed approach on the mini irrigation system.

Keywords Microcontroller · Dampness level · Moisture sensor · Smart irrigation

18.1 Introduction

Water saving is an important part of environment as the water largely affects the life of human [1]. The cheaper and easier availability of sensors strengthens the irrigation system by providing in-situ analysis of various sensor values including, humidity, moisture, temperature, etc. Such analysis estimates the requirement of water in terms of moisture level which helps to avoid the wastage of water. The

S. Chopade (✉)
K. J. Somaiya College of Engineering, Mumbai, India
e-mail: sandeepc@somaiya.edu

S. Chopade
Veermata Jijabai Technological Institute (VJTI), Mumbai, India
e-mail: schopade@mc.vjti.ac.in

S. Gawade
Pillai College of Engineering, New Panvel, India

A. Khosla and M. Aggarwal (eds.), *Smart Structures in Energy Infrastructure*,
Studies in Infrastructure and Control,
https://doi.org/10.1007/978-981-16-4744-4_18

requirement of accurate water amount can consume less water. It helps to identify exact water consumption for irrigation. In smart irrigation system, the sensors measure various characteristics of the soil and the generated data can be used to decide the amount of required water to the soil. The amount of water requirement can be identified by using different microcontrollers such as Arduino Uno. The automatic water requirement detection capability of microcontroller-based techniques enhances their use in irrigation system.

Figure 18.1 illustrates block diagram of a smart irrigation system by using moisture sensor. The sensory values are sent directly to the cloud. The moisture level of the soil is collected and given to microcontroller. Later, the data is also sent to the cloud which is made publicly available.

Further, a microcontroller-based approach is used to irrigate the soil if required depending on the moisture level. Finally, the moisture level is compared with preset values of microcontroller. Depending on the moisture level, the relay interface will start the solar-powered pump motor to water the plant. With the help of GSM module, this information is conveyed to the owner on his mobile phone using wireless medium.

The traditional techniques for irrigation system do not provide the mechanism to water the plant automatically. In other words, microcontroller-based technique can be applied to the irrigation system to water the plant as required without wastage of the water. In this work, we are using automatic irrigation technique; therefore, the microcontroller-based model can be applied to identify moisture level of the soil.

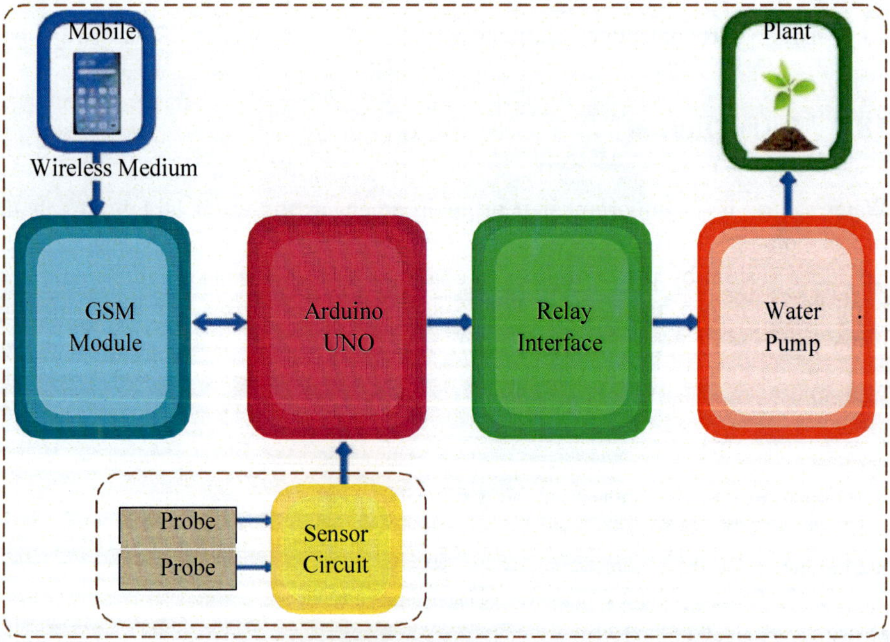

Fig. 18.1 Block diagram of smart irrigation system

18.1.1 Motivation

The work in this paper is motivated by the following major limitation in the existing literature:

- In existing work [2–4] on smart irrigation, microcontroller-based approaches are employed which requires GSM module for sending and receiving messages to and from mobile user. Such status of the soil moisture level is not available publicly on the Internet. The existing work [5] incorporates radio frequency (RF) module which makes the design of smart irrigation system very complex due to the sensitivity of radio circuits and accuracy of components. In [6], the smart irrigation system uses NRF24L01 to transfer the data over wireless media, we used ESP8266 to upload data to the cloud. In [7], Bluetooth is used for wireless communication which maximum covers 100 m distance and in [8], and Zigbee is used for wireless communication whose maximum coverage is 10 m only. In [9], algorithm is used to handle ON/OFF control operation of the pump motor which is time-consuming. In [10], the microcontroller is not used which is automatically handling whether to water the plant or not as well as the owner is not able to send or receive the message to or from the controller through wireless medium.
- The prior studies [7, 11, 12] on smart irrigation do not consider any solar-powered energy source but uses conventional energy sources which are having enormous problems including pollution and cost.
- The conventional irrigation system needs the manpower and wastage of water is observed, such as sprinklers, and flood-type system [13].
- The existing work [14] on smart irrigation adopted the mechanism to water the plant when intense requirement of water is there. It can harm the life of the plant if some delay for watering the plant will arise due to some problems, such as power failure and connectivity.
- The system [15, 16] focuses on how the crops can be prevented from the unconditional rain using fuzzy logic technique. The system is not explaining the use of any kind of non-conventional energy sources.

In this paper, we address the problem: How to use moisture level of the soil to automatically start the solar-powered pump motor using the interface between microcontroller and relay for irrigation purpose? To solve this problem, this work proposes a microcontroller-based approach that uses the concept of sensing the soil moisture. First, the proposed approach senses the moisture level of the soil. Next, we can interface Arduino Uno with moisture sensor to take the decision. Further, a message is conveyed to the mobile owner. Finally, depending on the status of the soil, the mobile user sends the message to start the motor pump or not. The interface between relay and Arduino Uno automatically starts the solar-powered pump motor without wastage of water.

18.1.2 Major Contributions

To the best of our knowledge, this is very essential work to address the problem of identifying moisture level in the soil using a microcontroller to automatically start the pump motor to water the plant if required. This paper makes the following major contributions:

- We propose a microcontroller-based approach that uses moisture sensor to identify the dampness level in the soil.
- This work proposes an algorithm that interfaces Arduino Uno with moisture sensor for deciding the moisture status of the soil.
- Next, this work also proposes an algorithm for automatic decision regarding the pump by comparing moisture value with preset values.
- Further, this work also proposes an algorithm to send and receive the message to and from mobile owner.
- Finally, we propose an interface between Arduino Uno and relay to actually start the solar-powered pump motor.

The rest of paper is organized as follows. Next section illustrates the terminologies and notations used in this work. Section 18.3 proposes microcontroller-based approach for identifying moisture level of the soil. In Section 18.4, it presents the experimental analysis of proposed approach. Finally, the paper is concluded in Section 18.5, with its future scope.

18.2 Preliminaries and Problem Statement

In this section, we first describe the different terminologies used in this work. Later, this section covers a brief description of the problem associated with irrigation system for plants and the overview of solution.

18.2.1 Preliminaries

The automatic irrigation system can be built by using the different microcontrollers including Arduino Uno, Zigbee, Raspberry Pi, etc. The building of irrigation system incorporates two methods, namely human operated and automatic sensor-based. In the human-operated method, the owner will see the moisture level in the soil, and after that according to the condition of soil, the owner will start the pump to water the plant if required. There may be the wastage of water. The human-operated method incorporates the presence of owner in person to check the condition of soil. If the owner is out of location then when he comes back, he will water the plant, which sometimes even two or more days. This delay in watering the plant hampers the real-time monitoring of the plant. The sensor-based automatic irrigation method gives a real-time monitoring of the moisture and water the plant without wastage of water. The Arduino Uno microcontroller is used for building an automatic irrigation system that can water the plant by starting the pump motor when the moisture level of the soil is not appropriate.

Definition 1 (*Irrigation System*). Irrigation system is a sophisticated system that supplies water to the land with the help of artificial canals, ditches, etc., for the promotion of different crops to increase the productivity.

Definition 2 (*Solar Tracking*). Solar tracking is a mechanism to gain maximum solar rays. The system automatically changes its direction to get the maximum sun rays, as there is decrease in the intensity of light.

Definition 3 (*Centrifugal Pump*). Centrifugal pump is a mechanical device which uses a rotating impeller to force a liquid fluid for moving forward inside a pipeline or hose. They also produce the pressure by the creation of a suction (partial vacuum), causing the fluid to rise to the higher altitude.

Definition 4 (*Impeller*). An Impeller is a rotating component of a centrifugal pump which transfers energy from the motor that drives the pump to the liquid fluid being pumped by accelerating the fluid outwards from the center of rotation. An impeller is usually a short cylinder with an open inlet (called an eye) to accept incoming fluid, vanes to push the fluid radially, and a splined or threaded bore to accept the motor driveshaft.

Definition 5 (*Casing*). The casing contains the liquid and acts as a pressure containment vessel that directs the flow of liquid in and out of the centrifugal pump. There are two basic types of pump casings: volutes and diffusers. The volute is a curved funnel that increases in area as it approaches the discharge port. On the other hand, a diffuser is a set of stationary vanes that surround the impeller.

18.2.2 Problem Statement and Overview of the Solution

Moisture level identification in the soil helps in protecting the crops by watering the plant at regular intervals without wasting the water resulting in increase of crop productivity, as discussed in the introduction. The moisture sensor data collected from the soil is used for watering the plant. The existing approaches are not capable enough to automatically water the plants. This work, therefore, addresses the problem of automatically starting the solar-powered pump motor by using only the moisture level.

Overview of the solution: This work proposes a microcontroller-based approach that uses the concept of automatic start of pump motor. The approach first uses a sensing the moisture value algorithm to check the moisture level of soil. Next, the approach sends the message to the mobile owner with the help of GSM. Later, the moisture value is used to start the pump motor by using relay automatically.

18.3 Proposed Approach

In this section, we propose design and implementation of solar-powered automated and smart irrigation model for public gardens, cricket stadium lawns, and golf lawns. The approach comprises four phases as follows, Phase 1: Sense the moisture level of soil, Phase 2: Taking decision whether to water the plant or not, Phase 3: With the help of GSM, convey message to the concerned authorized person, and Phase 4: Use of relay switch as interface between pump and microcontroller to control the operation of water pump motor. Phase 1 is a principal phase, as it incorporates sensing the moisture level of soil which is used for the next phase. Now, Phase 2 takes the decision automatically whether watering to the plant is needed or not. In Phase 3, a specific message is conveyed to the concerned authorized person with the help of GSM. The value of moisture level of soil informs the use of relay as interface between the pump and microcontroller as explained in Phase 4. Figure 18.2 shows all the phases involved in the proposed approach.

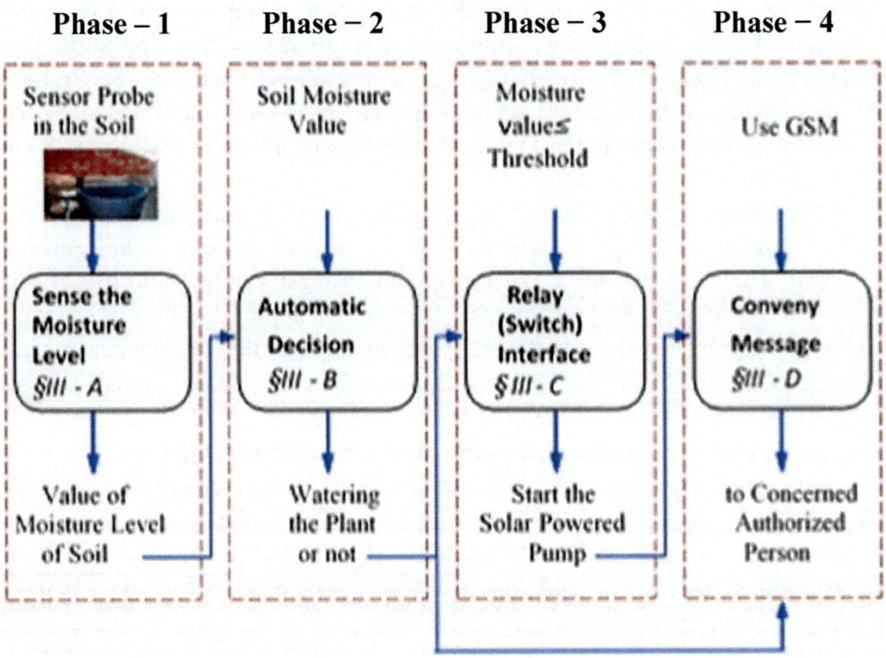

Fig. 18.2 Illustration of four phases involves in the proposed approach

(a) Tulsi plant (b) Moisture sensor (c) Sensor deployment.

Fig. 18.3 Different steps involved in sensing process

18.3.1 Sensing the Moisture Level of Soil

There exist a variety of sensors, such as temperature sensor for measuring the temperature of the soil, humidity sensor for checking the humidity of the soil, and nutrient sensor to measure the nutrients level in the soil. Here, we discuss a method for sensing the moisture level of the soil. During the sensing process, the soil moisture sensor probes placed in the plant pots inform the status of soil, such as dry, wet, or moist for the plants. With the help of this moisture sensor, the farmers can efficiently manage the irrigation system by using the measured quantity of water present in the soil. Thus, the farmers can save wastage of water by providing required water to the plant and increase the productivity as well as quality of the crop. Part (a), Part (b), and Part (c) of Fig. 18.3 illustrates one of the plants used in the experiment, the soil moisture sensor SEN92355P model having three pins utilized for ground, power, and flag, and deployment of the sensor into the test bed, respectively. Figure 18.3 shows that the soil moisture sensor checks the dampness of the soil.

This sensor reads the physical stimuli means the wetness of the soil and converts it into digital signal. Next, this digital signal is forwarded to the microcontroller for further processing.

18.3.2 Automatic Decision

Here, we interfaced a soil moisture sensor with the Arduino microcontroller. The soil moisture sensor incorporating two probes helps to measure the moisture of soil. The two probes are having the capability of passing the current through the soil which will result in the measure of the moisture value. We have connected the sensor in two modes: analog mode and digital mode. The microcontroller uses soil

(a) Arduino Uno (b) Connecting Arduino

Fig. 18.4 Connection between moisture sensor and Arduino Uno

moisture level data for automatic decision. The data collected by the microcontroller is used to compare with the preset values and then automatically deciding whether to provide water to the plants or not without actual being present at the location in person intervention. Figure 18.4 shows the microcontroller Arduino Uno board based on the specification Atmel ATmega328P. The board has 14-pin digital I/O, 6 analog inputs, 16 MHz ceramic resonance, USB connector, power connector, ICSP, and reset button. Next, it is connected to the computer using USB cable. As the microcontroller is ATmega16U2 (Atmega8U2 to R2), the driver is programmed as USB port converter instead of using FTDI USB serial chip. Arduino is the heart of the whole system.

In this work, the automatic watering to the plant is termed as automatic plant watering (APW) mechanism. The mechanism is advantageous by the fact that it only uses moisture content of the soil for starting the pump. First of all, we declare the object GSM and we pass the digital pin numbers as parameters. Here, pin number 2 will act as Rx of Arduino microcontroller and 3 will act as Tx of Arduino microcontroller. Let, sensor in represents the input pin on which analog input read from the sensor and connected to the analog input of Arduino. Output value variable stores the moisture value. In the configuration part, there exist the communication between the Arduino and serial monitor. The first task is to set baud rate to communicate with GSM module. The second task is to set the baud rate of Arduino Serial Monitor. Next, we set both baud rates as 9600 bits/s. This is the end of configuration part with setting baud rates and giving a small delay of 1 s. Next, we read the value from the sensor and store the values in the output value variable. Then, we map the output values to 0 100, because the moisture is measured in percentage. The sensor value in the dry soil was 1023, and in the wet soil, the sensor value was 470. So, we mapped these values to get the moisture content in the soil. The output value indicates whether the soil is dry or wet. Next, for dry soil the pump is started to water the plant by turning off the LED, and for wet soil, the message is sent to the mobile of not starting the pump by turning on the LED. The steps involved in APW mechanism are discussed in Procedure 1.

18.3.3 Send or Receive Messages Over GSM Module

This section presents the connection between the Arduino and the GSM module. In this work, we used the GSM module 900A capable of handling 2G, 3G, and 4G mobile networks for the connection of devices. Wi-Fi and Bluetooth are good and low-cost choices for but they work only at close ranges. When the devices are at remote locations, GSM is good and easy setup option. The GSM module is instructed through different AT commands to perform different work. AT means attention commands are special commands which follow UART protocol. There exist various AT commands for various functions.

Procedure 1. Automatic Plant Watering Mechanism

Input: Soil Moisture Value sensed by Moisture sensor
Output: Decision whether to start the Pump Motor or not
/* Declare object for GSM */
PIN 2 act R_x of Arduino and PIN 3 act T_x of Arduino
/* Soil sensor input at Analog PIN A0 */
sensorpin A0 ←
/* The setup function runs once */
void setup()
/* The loop function runs over and over again forever */
void loop()

Function *void setup()*
begin
/*Initialise communication between Arduino Uno & GSM*/
Set the baud rate to communicate with serial Monitor.
Initialize the digital PIN 12 as an output.
Turn the LED ON by setting HIGH voltage level.
Set baud rate to communicate with GSM module. Read from the sensor.
Set baud rate to communicate with GSM module.
Set the baud rate to communicate with serial Monitor.

Function *void loop()*
begin
Turn the LED ON by setting HIGH voltage level.
Convert analog input to a digital number between 0 to 1023.
Map this value between 0 to 100 percent.
if $value \leq 90$ **then**
Set the GSM module in Text mode.
Send the message Moisture Low to Mobile phone.
The end of SMS CTRL+Z symbol.
Print SMS sent.
Turn the LED off by making the voltage LOW.
Send the message Pump Motor ON.
Turn the LED ON by setting HIGH voltage level.
if value >90 **then**
Turn the LED ON by setting HIGH voltage level.
Send the message Pump Motor OFF.

At the beginning, blinking rate of network LED is very fast, indicating the GSM module is still searching the network. Part (a) and Part (b) of Fig. 18.5 illustrate the SIM 900A GSM module with antenna and connection with Arduino Uno in our work.

The Arduino Uno board used in the previous section can be utilized to send and receive messages by connecting GSM module with it. In this work, the GSM module uses the same setup which is already incorporated in the previous section for completing the configuration part. After, we checked whether the data is available on GSM or not. If it is available then receive or read on the Arduino Uno board. Next, we handled two different cases depending on the value of data. In the first case, Arduino Uno sets the GSM module in text mode using AT commands. Next, the Arduino Uno board is connected to the personal computer. With the help of computer, moisture value sensed by the sensor is sent to the mobile phone. Further, in the second case, the LED is low indicating the start of motor is sent to the mobile user. Procedure 2

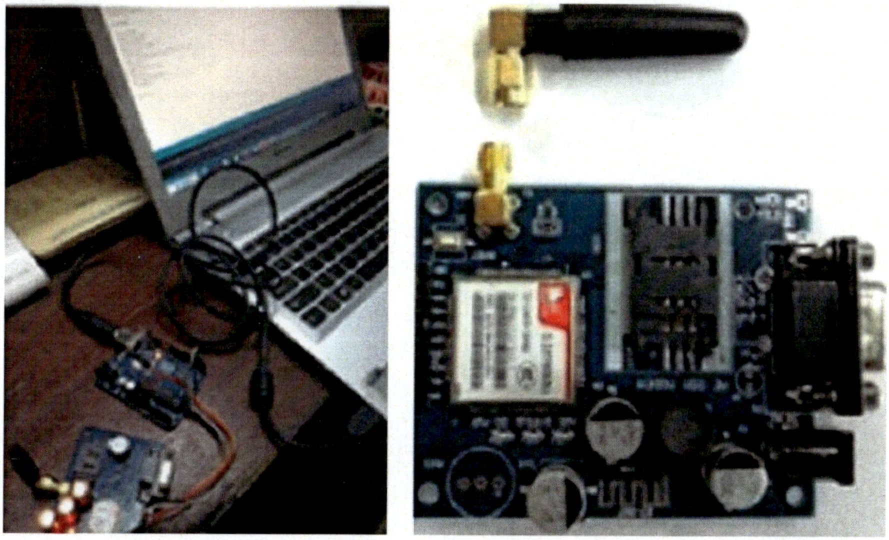

(a) GSM (b) Set up of GSM

Fig. 18.5 Connection between Arduino and GSM module

illustrates all the steps involved in the sending and receiving messages to and from the mobile phone using GSM module. The setup() function is similar as discussed in Procedure 1.

Procedure 2. Send and Receive Messages

Input: AT Commands to instruct GSM Module
Output: Send and receive messages to and from GSM
/* Declare object for GSM */
PIN 2 act as R_x of Arduino Uno and PIN 3 act as T_x of Arduino Uno
/* The setup function runs once */
void setup()
/* The loop function runs over and over again forever */
void loop()

Function *void loop()*
begin
/* Arduino instructs GSM module */
if (Data is available on GSM) **then**
Receive or read a live SMS on Arduino Uno.
if (Data on GSM is positive) **then**
Write data on Arduino Uno serial monitor.
switch (Data from Arduino Uno)
Case 1:
Arduino Uno sets GSM module in the Text mode.
Use Mobile number to send SMS.
Send Moisture value to Mobile phone.
Send end of Message.
Case 2:
Turn LED OFF (Low Voltage) on Arduino Uno.
Send the Motor ON SMS to Mobile phone.

18.3.4 Relay Interface with Arduino Uno

In this section, we interface the relay with the Arduino Uno to actually start or not to start the solar-powered pump. Here, Arduino Uno provides 5 V DC power supply to control the Relay. When the moisture level in the soil is low, then first the relay will be tuned on and then the pump motor. Similarly, when the moisture level is high, then Relay will be turned off, in turn, cutting down the power supply to the pump motor. Procedure 3 illustrates all the steps involved in the process of starting or not starting the solar-powered pump motor.

Part (a) of Fig. 18.6 shows relay used in our work, respectively. Here, we used relay switch to control the operation of motor and we utilized 775 motor capable to provide large torque, high power with very less noise, and rotary motion to a pump. It is a brush-commutated DC motor having various applications, namely power tools, water pump, household appliances, fan, etc. Part (b), Part (c), and Part (d) of Fig. 18.6 illustrates pump motor used in our work, Bio-Medical Engineering and Technology Incubation Cell (BETIC) laboratory of our institute, where we produced a pump with the help of 3D printing machine utilized in our work, and actual interfacing of relay with water pump motor, respectively.

Procedure 3. Relay interfacing with Arduino Uno

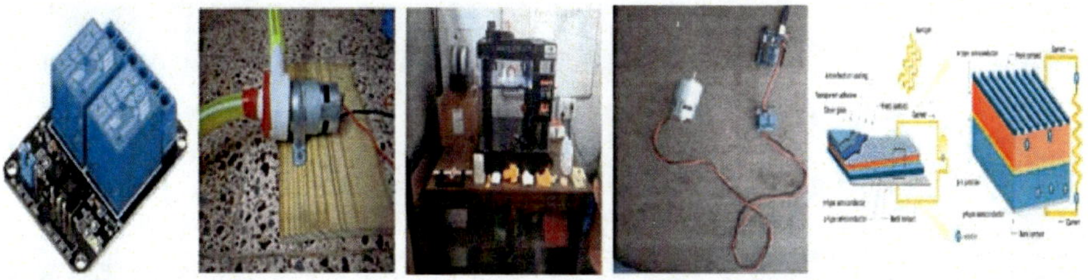

(a) Relay (b) Pump. (c) BETIC. (d) Connection of Relay Pump. (e) Solarcell structure & Working

Fig. 18.6 Various steps involved in relay interface

Input: 5V DC power supply to Relay from Arduino Uno
Output: Pump Motor ON or OFF
/*The setup function runs once*/
void setup()
/* The function runs over and over repeatedely forever */
void loop()

Function *void setup()*
begin
Initialize the digital pin 12 of Arduino Uno in output mode.

Function *void loop()*
begin
if(*moisture* $\leq$ 90)**then**
Turn LED OFF by making voltage LOW.
Relay switchON.
Pump MotorON.
if (*moisture* >90) **then**
Turn LED ON by making voltage HIGH.
Relay switchOFF.
Pump MotorOFF.

In the proposed work, water pump motor as well as the whole project is powered by the solar energy generated by the solar panel converting sunrays into electrical energy which is stored in the battery and battery is used to operate the pump. Part (a) of Fig. 18.7 illustrates the lead–acid battery used in our work, as it is rechargeable, fully sealed, and commonly used in backup power sources such as uninterruptable power supply units having 12 V 4 Ampere output. Again, this battery is generally used in bikes and motorcycles also. Next, the solar charge controller manages the power going into the battery with the help of solar array of solar panel as shown in the Part (b) and Part (c) of Fig. 18.7.

The solar charge controller gives assurance that the deep cycle batteries are not overcharged during the daytime. Further, it ensures that the power does not run back to solar panels overnight and drain out the batteries. The solar panel is capable to provide maximum DC voltage of 16.8 V which the solar charge controller converts into DC voltage of 12 V for charging the battery. Further, we used the centrifugal

(a) Battery. (b) Solar panel. (c) Charge Controller

Fig. 18.7 Various steps for battery

pump which is designed and manufactured on 3D printing machine in our BETIC laboratory. It is suitable for the transfer of low viscosity fluids, namely water, etc., in high flow rate, low-pressure installations, making them ideal for large volume applications. Moreover, the construction of a centrifugal pump is so simple, making them easy to manufacture in various materials, such as plastics, and cast iron for lighter duties applications. Thus, we used the plastic material for manufacturing a centrifugal pump which is utilized in our work. Further, we used the volute casing of a centrifugal pump which slows down the received fluid's rate of flow. Moreover, we proposed the solar panel efficiency by calculating the optimal tilt angle using simulation softwares, such as RETScreen and PVSYST. Afterward, we decided that 19° toward south is the best angle of tilt for solar panel placed on terrace of Renewable energy laboratory, Mechanical Engineering Department, third floor, K.J. Somaiya College of Engineering, Mumbai.

18.4 Experiments and Results

In this section, we first discuss the implementation details of the proposed approach. Later, the approach is evaluated, under the experimental setup. Finally, this section presents a comparative analysis of the proposed approach with the existing work.

18.4.1 Implementation Details

This section discusses the implementation of the proposed approach, which follows similar phases, as mentioned in Section 3. We implemented the proposed approach in Arduino Uno IDE, open-source software. SolidWorks software used to create 3D design model of Impeller and Casing of pump along with the assembly of motor and pump in BETIC laboratory of our institute. We used ANSYS software for analysis of different design steps of pump and RETscreen and PVSYST softwares for solar panel.

The implementation of four phases are as follows:

(1) The implementation of the first phase incorporates a hardware connection for interfacing Arduino Uno with soil moisture sensor. This sensed soil moisture value is used to decide whether to start the pump or not in Phase 2.
(2) The second phase is implemented in Arduino Uno IDE.
(3) In Phase 3, sending and receiving messages to and from the mobile user incorporates the use of GSM module.

The interfacing between GSM module and Arduino Uno is implemented in Arduino Uno IDE.

(4) Next, the relay interfacing with Arduino Uno is implemented again in Arduino Uno IDE to start the solar-powered pump.
(5) The design of the pump is done with the SolidWorks software and it is manufactured in BETIC laboratory.
(6) Analysis of the pump to test the design of the pump is done under various stress conditions in ANSYS software.
(7) Finally, solar panel efficiency is achieved by doing simulations in RETscreen and PVSYST softwares.

18.4.2 Experimental Results

In this section, several experiments are carried out to design and evaluate the performance of the proposed approach and provide answers to the following questions:

- What is the criteria for the selection of pump? (Section IV-B1)
- How does the selected pump designed and manufactured? (Section IV-B2)
- What is the criteria for the selection of motor? (Section IV-B3)
- How does the selected pump analyzed? (Section IV-B4)
- How does the different phases of automatic irrigation system assembled? (Section IV-B5)
- How to build a mini irrigation system? (Section IV-B6)

Table 18.1 Comparison of pumps used in our experimental setup

Factor	Centrifugal pump	Positive displacement pump
Construction	Simple	Complicated
Design	Less moving parts	More moving parts
Viscosity of fluid	Low viscosity fluids	Low and high viscosity fluids
Flow rate	High	Low
Pressure	Low	High
Maintenance	Lower maintenance	Higher maintenance
Cost	Cheaper	Expensive

(1) Selection of pump: The experiment starts with the selection of pump. We evaluated our experimental setup by utilizing different pumps by considering various factors of the pump and arrived at the decision of using a centrifugal pump, due to the observed comparison as shown in Table 18.1. The centrifugal pump has given much better results for our setup.

(2) *Design and manufacturing of selected pump*: The impeller and casing of the selected centrifugal pump are designed in SolidWorks software, and it is manufactured in BETIC laboratory of our institute discussed in Section 3. We have designed and used the volute casing as it is the curved funnel that increases in area when it approaches toward the delivery port. Part (a) of Fig. 18.8 illustrates the impeller component having inlet vane angle 10° and outlet vane angle 45° incorporated in our work. Next, a casing which is a curved funnel having

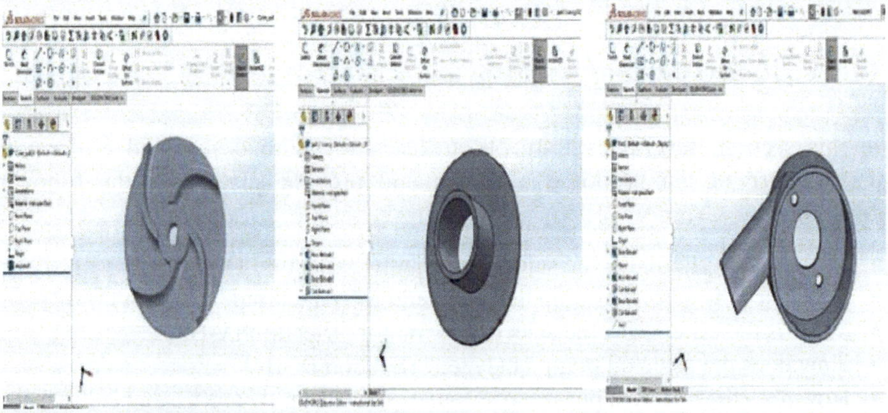

(a) Impeller (b) Suction casing (c) Delivery casing

Fig. 18.8 Different parts of pump used in our work

Table 18.2 Various features and their dimensions used in our work

Feature	Dimensions (mm)	Feature	Dimensions
Shaft diameter	5	Shaft size	17 mm
Body length	66.7	Front steps diameter	17.4 mm
Former high level	4.7	Body diameter	42 mm
Motor length	98	Mounting hole size	M4
Mounting hole	2	Torque	2 kg cm

external diameter 54 mm, thickness 2 mm, and internal diameter 20 mm as shown in Part (b) and Part (c) of Fig. 18.8 utilized in our work.

(3) *Selection of motor*: Further, among the various motors tested for our experiment, 775 motor is best suited for our work having physical specifications as shown in Table 18.2. Table 18.3 summaries various speeds in RPM of 775 motor under various currents and voltages incorporated in our work. Part (a) and Part (b) of Fig. 18.9 illustrates the 775 motor with dimensions and with various component labels used in our work, respectively. Assembling of the impeller

Table 18.3 Speeds of 775 motor under various input conditions

Voltage	Current	Speed	Voltage	Current	Speed
12	0.14	3500	18	0.15	4500
24	0.16	7000	30	0.17	8100

(a) 775 Motor with dimensions

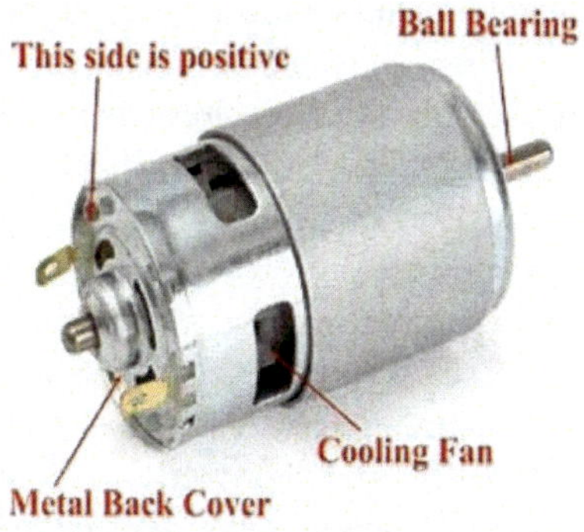

(b) 775 Motor with component labels

Fig. 18.9 775 motor used in our work

Fig. 18.10 Assembly of pump

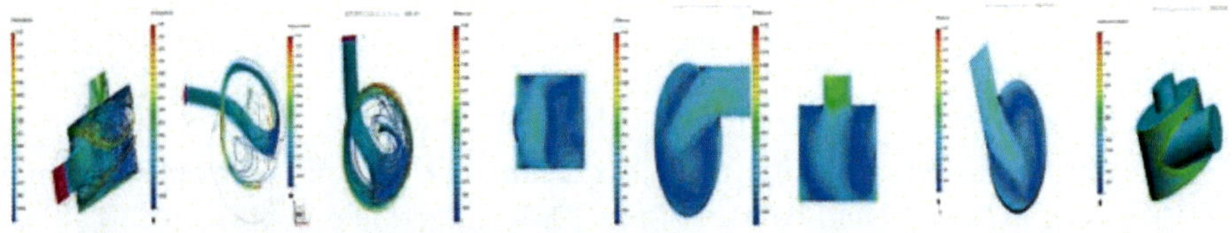

Fig. 18.11 Analysis of pump using ANSYS software

casing of the used centrifugal pump and 775 motor is shown in Fig. 18.10 for further work. We generated the assembly using SolidWorks software.

(4) *Analysis of Centrifugal pump*: We have analyzed the centrifugal pump under input conditions, such as inlet velocity 5 m/s and motor speed 5000 RPM with the help of ANSYS software. We found that our design of the pump is safe under wall shear stress. Figure 18.11 illustrates the analysis of a centrifugal pump by using ANSYS software with various attributes tested, namely velocity and wall shear stress.

(5) *Assembly of all subsystems*: After procurement of all materials, electric components, and manufacturing of designed pump, different subsystems are assembled and tested individually. Initially, the moisture sensor is tested whether it is reading the moisture level successfully or not by interfacing Arduino Uno with the sensor and then the moisture value displayed on the computer monitor. After immersing the moisture sensor in the dry soil, the reading obtained displayed was 1023. Thus, 1023 moisture value indicates zero moisture level in the water. Next, after pouring water in the soil, the sensor again immersed in it. This time, the moisture value being displayed started falling down and reached up to 361, indicating the maximum moisture. Further, we tested the interfacing between Arduino Uno and GSM by checking whether the module is sending and receiving the messages to and from the end user as well as Arduino Uno able to read and properly respond to the messages received from module. The GSM module is tested for the operations, such as sending the message, calling the user, and receiving a text message. Finally, we tested the interfacing between Arduino Uno and Relay. When the moisture is low, the relay started the pump motor to water the plant. The battery which is charged by solar panel is connected to motor to provide the power supply. Figure 18.12 shows

Fig. 18.12 Complete irrigation system setup

the creation and assembling of various subsystems in corporate in automatic irrigation system.

(6) *Building a mini irrigation system*: In this section, we have tested out the working model made up of various materials, such as PVC pipes, wooden strips of size 1 cm and 1.5 cm, screws of having U joints and L joints, soil having pH level of 6–7.5, at least 4 h of sunlight daily to generate solar power, watering the plant when top 1 inch of the soil is dry, and tulsi plant. Figure 18.13 shows support column used for mini irrigation system. Table 18.4 shows the number of times the particular component is switched ON in that specific day. The moisture sensor immersed in the soil detects the low moisture level and sends a message to the mobile user. Further, the mobile user sends the message of MOTOR ON and the motor starts. Due to some technical issues, such as battery being

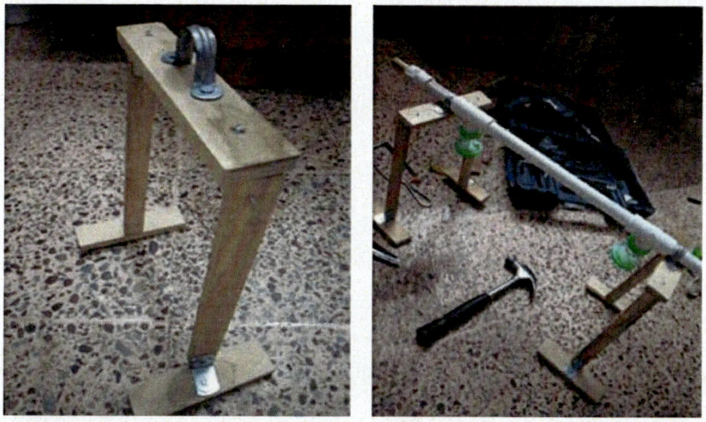

Fig. 18.13 Support column used in mini irrigation system

Table 18.4 Readings of a mini irrigation system

Date	Soil moisture low	Message received	Motor ON
21/04/19	3	3	3
22/04/19	3	2	2
23/04/19	4	4	4
24/04/19	4	4	4
25/04/19	3	3	3
26/04/19	3	3	2
27/04/19	3	3	3

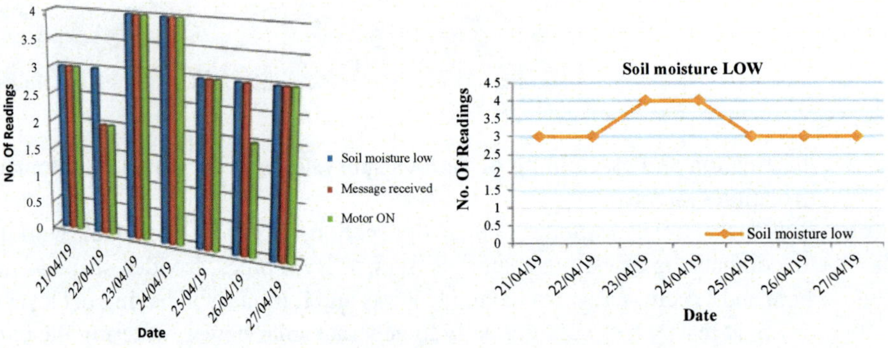

(a)(b) Soil Moisture Low information

Fig. 18.14 Soil moisture low information during working of mini irrigation system

undercharged or the mobile phone being out of network coverage, on some days we do not get a message since the soil moisture level was low.

Part (a) of Fig. 18.14 illustrates the observations of how many times the pump has been started to water the plant on a particular day based on the moisture level information transmitted by the sensor and Part (b) of Fig. 18.14 shows the number of times the watering to the plant required due to low soil moisture level. Part (a) of Fig. 18.15 illustrates the observations of how many times the owner received the message and Part (b) of Fig. 18.15.

18.5 Conclusion

In this paper, we proposed a smart and automatic irrigation system by reducing human intervention for growing plants. We utilized solar energy to run the system and automate it with the help of various electric components. We also proposed solar panel efficiency by analyzing various types of methods and achieved efficiency by calculating optimal tilt angle. This proposed system works successfully in the

Fig. 18.15 Message received and motor ON information during working mini irrigation system

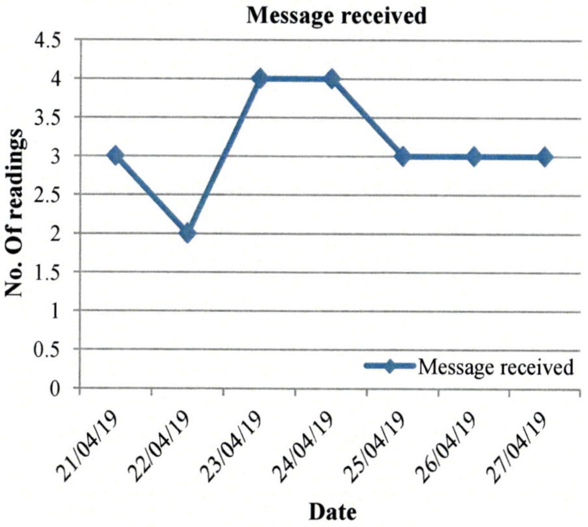

(a) Message received information

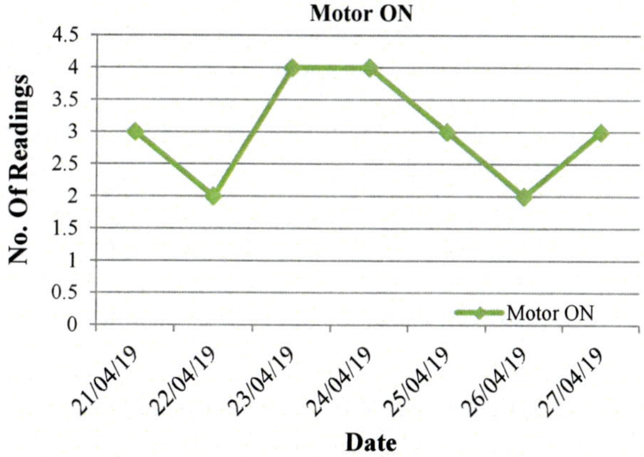

(b) Motor ON information

absence of the human intervention by finding the condition of the soil through a moisture sensor. Next, according to the condition of the soil, Arduino Uno, heart of the system, runs the irrigation system. It also gives information to its mobile user about the condition of soil and pump motor. Here, the availability of the electricity is not mandatory, as solar power is the major source of the electricity to start the pump motor. Due to reasonable cost, this method of watering to the plants without wastage of water is useful for farmers particularly in rural areas, public gardens in the towns. During the performance testing of our built mini irrigation system for terrace, we achieved maximum automation as per our expectation level without wastage of water. We also tested that; the proposed system ran the pump motor without sunlight by using stored energy in the batteries. This work also provides a further research direction toward incorporating humidity, temperature, and other sensors to increase the productivity of crops. Our model is a small-scale model which is useful for terrace or balcony gardens. We can scale up its which will be useful for farmers where there will be no availability of regular electricity.

Acknowledgements In this section, we would like to acknowledge the continuous motivation, support from the management, principal, and department for the innovative practices. We would like to thank them for providing such a nice renewable energy lab on the third floor. We would also like to acknowledge the participation of our students and faculty who has directly or indirectly contributed in this research and in the execution of the various activities related to it.

References

1. J. Dong, G. Wang, H. Yan, J. Xu, X. Zhang, A survey of smart water quality monitoring system. Environ. Sci. Pollut. Res. **22**(7), 4893–4906 (2015)
2. S. Suman, S. Kumar, R. Sarkar, G. Ghosh, Solar powered automatic irrigation system on sensing moisture content using Arduino and GSM. Int. J. Adv. Res. Electron. Commun. Eng. (IJARECE) **6** (2017)
3. E. Ososanya, S. Haghani, W. Mahmoud, S. Lakeou, S. Diarra, Design and implementation of a solar-powered smart irrigation system, in *American Society for Engineering Education, ASEE Conference Paper ID*, vol. 13224 (2015)
4. S.E. Babaa, S.A. Al-Jahdhami, M. Ahmed, S.A. Khan, B.S. Ogunleye, J.R. Pillai, Smart irrigation system using Arduino with solar power
5. D. Rane, P. Indurkar, D. Khatri, Review paper based on automatic irrigation system based on RF module. Int. J. Adv. Inf. Commun. Technol. **1**(9), 736–738 (2015)
6. P. Rajalakshmi, S.D. Mahalakshmi, IoT based crop-field monitoring and irrigation automation, in *2016 10th International Conference on Intelligent Systems and Control (ISCO)* (IEEE, 2016), pp. 1–6.
7. M. Monica, B. Yeshika, G. Abhishek, H. Sanjay, S. Dasiga, oT based control and automation of smart irrigation system: an automated irrigation system using sensors, GSM, Bluetooth and cloud technology, in *2017 International Conference on Recent Innovations in Signal processing and Embedded Systems (RISE)* (IEEE, 2017), pp. 601–607
8. S.B. Saraf, D.H. Gawali, IoT based smart irrigation monitoring and controlling system, in *2017 2nd IEEE International Conference on Recent Trends in Electronics, Information & Communication Technology (RTEICT)* (IEEE, 2017), pp. 815–819
9. S.S. Sheikh, A. Javed, M. Anas, F. Ahmed, Solar based smart irrigation system using PID controller, in *ICAET-2018*, IOP Conf. Ser.: Mater. Sci. Eng. **414**(1), 1–8 (2018)
10. S. Harishankar, R.S. Kumar, K. Sudharsan, U. Vignesh, T. Viveknath, Solar powered smart irrigation system. Adv. Electron. Electr. Eng. **4**(4), 341–346 (2014)
11. S. Malge, K. Bhole, Novel, low cost remotely operated smart irrigation system, in *2015 International Conference on Industrial Instrumentation and Control (ICIC)* (IEEE, 2015), pp. 1501–1505
12. H. Benyezza, M. Bouhedda, K. Djellout, A. Saidi, Smart irrigation system based thingspeak and Arduino, in *2018 International Conference on Applied Smart Systems (ICASS)* (IEEE, 2018), pp. 1–4
13. Guidelines for Crop Production, Management, and Irrigation (2019) [Online]. Available at: https://www.toppr.com/guides/biology/crop-productionandmanagement/irrigation/
14. R. Suresh, S. Gopinath, K. Govindaraju, T. Devika, N.S. Vanitha, Gsm based automated irrigation control using raingun irrigation system. Int. J. Adv. Res. Comput. Commun. Eng. **3**(2), 5654–5657 (2014)

15. R. Santhana Krishnan, E. Golden Julie, Y. Harold Robinson, S. Raja, R. Kumar, P.H. Thong, L.H. Son, Fuzzy logic based smart irrigation system using internet of things. J. Clean. Prod. **252**, 119902 (2020). ISSN 0959-6526
16. S.M. Ahmed, B. Kovela, V.K. Gunjan, Solar-powered smart agriculture and irrigation monitoring/control system over cloud—an efficient and eco-friendly method for effective crop production by farmers in rural India, in *Proceedings of International Conference on Recent Trends in Machine Learning, IoT, Smart Cities and Applications*. Advances in Intelligent Systems and Computing, vol. 1245, ed. by V.K. Gunjan, J.M. Zurada (Springer, Singapore, 2021). https://doi.org/10.1007/978-981-15-7234-0_24

Chapter 19
Drone Development and Embellishing It into Crop Monitoring and Protection Along with Pesticide Spraying Mechanism

Smita Agrawal, Preeti Kathiria, Vishwam Rawal, and Trushit Vyas

Abstract Agriculture is one of the most vital industries for a developing country. As far as India is concerned, over 70% of rural people rely on the production of crops for agriculture. However, they do face huge amounts of loss due to disease spreading. Naturally, it happens due to pests and insects that revive around the field, which decreases the productivity of crops. For averting the bugs and pests, several pesticides and fertilizers have been used to increase the productivity of crops. It is very much difficult and dreadful indeed for farmers to spray the pesticides on crops manually. World Health Organization (WHO) extrapolated about 10 lakh cases of ailing from this disease are affected because of spraying the pesticides manually. Drone aircraft is one of the proper remedies to avert this disease. Rather than spraying it manually, UAV sprays it automatically, which also deduce the health problems that farmers are facing. This paper conveys the brief deployment of UAV for crop overlooking and chemical spraying along with different UAV types and also discussed research opportunities in various domain.

Keywords Agricultural productivity · Unmanned aerial vehicle (UAV) · Pest detection · Spraying system · Crop monitoring

S. Agrawal (✉) · P. Kathiria · V. Rawal · T. Vyas
Nirma University, Ahmedabad, India
e-mail: smita.agrwal@nirmauni.ac.in

P. Kathiria
e-mail: preeti.kathiria@nirmauni.ac.in

V. Rawal
e-mail: 18bce374@nirmauni.ac.in

T. Vyas
e-mail: 18bce376@nirmauni.ac.in

A. Khosla and M. Aggarwal (eds.), *Smart Structures in Energy Infrastructure*,
Studies in Infrastructure and Control,
https://doi.org/10.1007/978-981-16-4744-4_19

19.1 Introduction

One of the main sources of Indian financial system is agriculture. Agriculture industry relies upon a variety of environmental parameters like temperature, rain, etc. [1]. In India, 70% of people are farmers, and they do not have colossal farming land due to little parts of land, and significant of the agriculture task like scattering, plowing, cultivating, and chemical spraying has done manually. And spraying chemicals are typically used to perish bugs, which helps to maximize crop productivity. According to the WHO, more than 10 lakh cases of unwell health have seen because pesticides recorded, and among them close to about 10 lakh, human beings are tussling from major illness problems. These chemical fertilizers distress quickly in the human vein system and can result in a severe impact on the body. These pesticides cause so many health problems, which also additionally reason cancer [2]. This paper signifies us about unmanned aerial vehicle (UAV) with a spraying mechanism that helps the agriculture industry for pesticide spraying in a proper manner at a low cost. This paper describes unique types of drones with their type and specific parameters like mass, area of flying, payload configuration, and their price.

19.2 Unmanned Aerial Vehicle (UAV)

Unmanned aerial vehicle (UAV) is capable of flying without a pilot [3]. Drone can be managed dynamically either via radio waves or autonomously (i.e., constant route). It is not necessary to have a particular size or kind of drive for a drone. They are often equipped with accessories used for surveillance and monitoring, in the shape of the optoelectronic heads. The most necessary function of the drones is that they do not require any extra infrastructure to quickly register and monitor a particular location or object [3]. Different varieties of drone (UAV) design used till now are shown in Fig. 19.1 also summarized the various UAV models used for precision agriculture in Table 19.1. Fixed wing Fig. 19.1a drone has one inflexible wing, and it is designed to look and work like an airplane. This type of drone is no longer capable of standing or flying in the same place. Single rotor Fig. 19.1b drone is robust and looks like a helicopter in design and structure. It has only a single huge rotor, which is like big

Fig. 19.1 UAV types, embedded wing (**a**). Single rotor (**b**). Quad copter (**c**). Hexa copter (**d**). Octo copter (**e**)

Table 19.1 Several UAV models used for precision agriculture

Type	References
Quad copter	[1, 2, 4–9]
Hexa copter	[5, 10–13]
Octo copter	[5, 14–16]
Single rotor	[17–19]

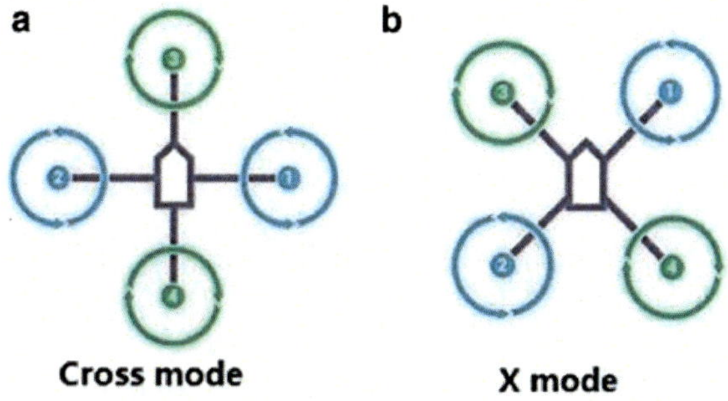

Fig. 19.2 Quad copter—**a** + configuration. **b** X configuration

spinning wings as helicopters have. It also has a small dimension rotor on the path for direction and balance. Quad copter (Fig. 19.1c), Hexa copter (Fig. 19.1d), and Octo copters (Fig. 19.1e) are multirotors that are uplifted and propelled by 4, 6, 8 rotors.

A quad copter is a multicopter that is propelled and lifted by four propellers (rotors) [4]. Two motors rotate clockwise (CW) direction, and other two rotate counter clockwise (CCW) direction to create opposite force to stay balance. The quad copter motion round the axis consists of backward pitch, forward pitch, left roll, right roll, clockwise yaw, and counterclockwise yaw [5]. It is the most famous multirotor with the easiest configuration. Configurations like cross (x) model are shown in Fig. 19.2b and plus (+) model shown in Fig. 19.2a.

19.3 Methodology

The flight controller is the brain of UAV that is used to perform main operations of drones like fly up, down, rotate, and other operations [3], and the control systems are equipped with the akin set of sensors with the difference in the velocity interpolation and in several algorithm methodology used.

The block diagram of UAV model represents in Fig. 19.3 which shows the interfacing of various components to design drone. For better understanding, various strategies according to requirement parameters like speed and load reviewed the controller used by researcher in UAV for agriculture domain represents in Table 19.2. Electronic speed control (ESC) mainly focuses on motor rpm [3]. UAV signals are sent through the radio channel transceiver. Every RC transmitter has a variety of

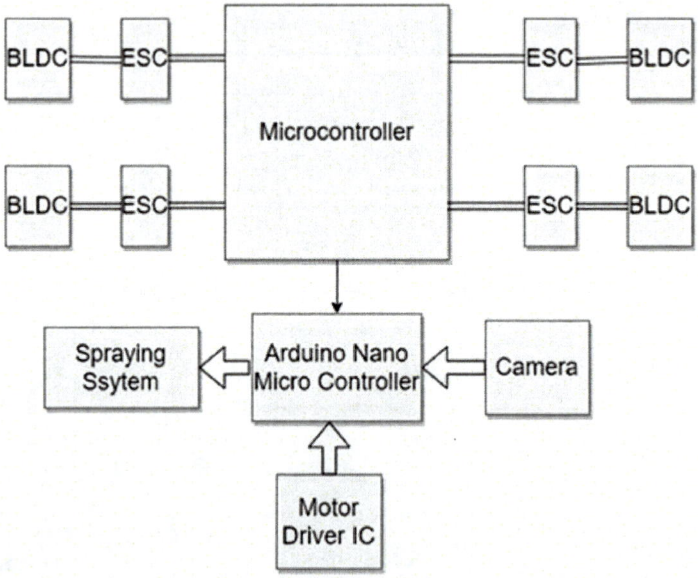

Fig. 19.3 Block diagram of model UAV

Table 19.2 Different controllers and methodologies

Article	Controller	Strategy	Load	Speed (rpm)
[4]	Flight controller KK 5.5 ATMega	AR 700 wireless receiver	1 kg	1000
[1, 9]	Flight controller Pixhawk 2.4.8	WSN	1.8 kg	1000
[1, 8]	Avionic RCB OS-10 Transceiver	RC transceiver	2.4 GHz	–
[2]	ATMega 2560	Radio receiver	2 kg	1000
[6]	Pixhawk 4	WSN, RC transceiver	1.5 kg	1000
[6]	Skyroid RC and Camera	RC transceiver	40 g	–
[7]	ATmega328	Radio receiver	–	1000
[8]	ArduPilot	Radio receiver	–	–
[11]	Arduino	AR 700 wireless receiver	2–3 kg	–

channels for solitary activity to get the UAV in control [5]. Power to drone is supplied through batteries which is the main disadvantage because it gets over after around fifteen minutes of flight. Primary cells are the ones which cannot be recharged and have to be removed after the expiration of the lifetime, whereas secondary cells are required to be recharged when the charge receives over. Both the kinds of battery are used significantly in different types of drones, and these cells range in size and material used in them. But mostly lithium polymer (LiPo) batteries are used in drones which is one type of secondary cell [3].

19.3.1 Sprinkling System

On the down side of the drone, which has a nozzle below the pesticide tank, the sprinkling device is attached to sprinkle the chemical closer to the down side. The sprinkling system is separated into two different portions, one is the sprinkling system itself, and the next one is the main controller board which represents using block diagram in Fig. 19.4. The sprinkling system carries the spraying components like pests or chemicals. And, it also includes a nozzle for pest spray [5]. The mechanism for spraying basically consists of a microcontroller which is programmed to work some of the functions. In the sprinkling system, different components are installed, and liquid tank capacity of 1 L and a pump is connected to a delimiter that spits the chemical the number of nozzles. It also includes a hardware component like a motor driver circuit that controls the velocity of spraying chemicals. Tank stage indicator circuit with alarm is there for recommends the pesticides is empty in tank and need to refill [1].

The spraying methodology executes the following instructions:

- **Pump ON/OFF control**: It performs the activation and deactivation of water spraying pumps.
- **Spraying Speed Control**: This is used to control the spraying speed which is done through sending signals to the brushed ESC motor.

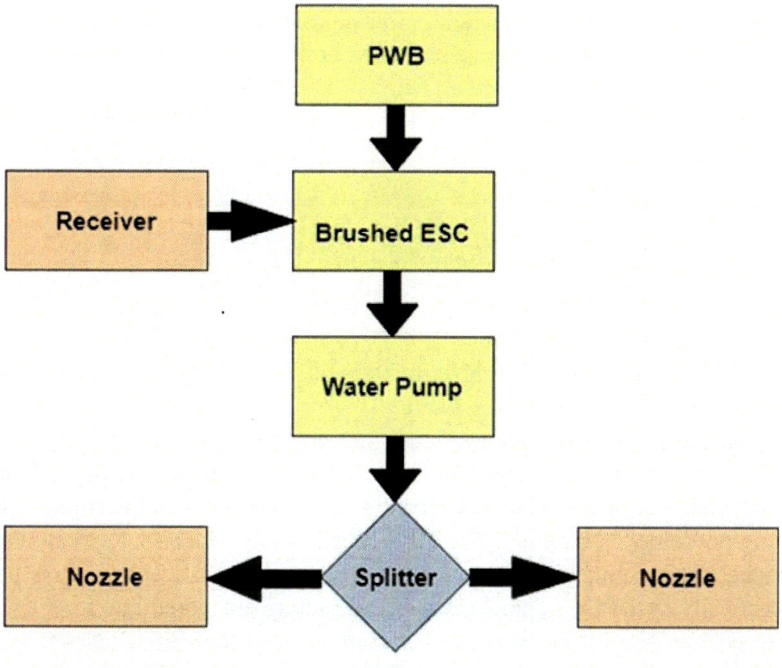

Fig. 19.4 Block diagram for spraying system

- **Tank status**: This is used to monitor the level of the tank by using a water degree sensor. If the pest in the pesticide tank goes below the threshold level, and then it will notify through a buzzer or alarm. Hence, the buzzer or alarm alerts the operator, and then he/she can land the drone for refilling the pest.

19.3.2 Hardware Requirement for UAV

For its stirring control in accordance with the sensed conditions, there are a range of components fixed in the UAV. In addition, larger elements that threaten the UAV, including specific sensors, characteristics, and points, are also embellished. Different kinds of hardware components and peripherals devices which are used in UAVs are listed in Table 19.3.

19.4 Crop Monitoring

UAVs are capable of tracking the crop with different parameter types and are capable of covering hectares of field in a single flight. And thermal and multispectral cameras mounted at the bottom of the drone are used to track crops. The camera takes one capture per second and stores it in memory and sends it to the application of the receiver and uses protocol for wireless transmission.

Table 19.3 shows the components which are used for drone building, and each component purpose is mentioned above.

The information collected through telemetry from a multispectral digital camera was analyzed using Eq. (19.1) represented normalized difference vegetation index (NDVI) geographic indicator.

$$\text{NDVI} = (R_{\text{INR}} - R_{\text{RED}})/(R_{\text{INR}} + R_{\text{RED}}) \tag{19.1}$$

where

R_{INR} = Reflectance of the near-infrared band,

R_{RED} = Reflectance of the red band.

The resulting calculation shows values ranging from -1 to $+1$; the value closer to 0 (ZERO) shows no crop foliage, and the value closer to $+1$ shows the full thickness of the green leaves on the crop. Based on the results, farmers can easily identify where to spray pesticides in the field. The GPS coordinates of each picture captured from the camera are maintained via an inbuilt GPS module. The UAV sprays pesticide automatically as the GPS coordinates of that picture are stored into it.

Table 19.3 Different equipment's and peripherals devices used in UAV for precision agriculture

Reference	Components	Purpose
Accelerometer	[1] [5] [10]	Measure linear acceleration based on vibration
Gyroscope	[1] [5] [10]	Determine the angular position
BLDC motor	[1] [2] [3] [4] [5] [6] [7] [8] [9] [10]	It works on speed of 1000 rpm
Propeller	[1] [2] [3] [4] [5] [6] [7] [8] [9] [10]	Propeller wings are used to lift
Brushless ESC	[1] [2] [3] [4] [5] [6] [7] [8] [9] [10]	To spin all motors on same speed
Brushed ESC	[2]	Used to control water pump
KK 5.5	[1]	Microcontroller/flight controller
AR 700 receiver	[1]	It has 7 channels which allows 7 commands to be sent at once
Lipo 3500/5200/5000/5500 mAh	[1] [2] [3] [4] [5] [6] [7] [9] [10]	LiPo cells have a nominal voltage of 3.7 V
12 V DC submersible water pump	[2] [3] [7] [8] [9] [10]	Water pump used for spraying pest
Nozzle	[2] [3] [7] [9] [10]	For spinking pest
Pixhawk 2.4.8	[2] [9]	Microcontroller
Avionic RCB OS-10 transmitter	[2]	Device which translates the pilot command in movement of motor
Avionic RCB OS-10 receiver	[2]	Used to receive signal transmitted from transmitter
IMAX B6 charger	[2]	Charger used to charge battery
MPU 6050	[3]	Sensor combination of gyroscope and accelerometer
GPS	[3] [6]	For drone location
ATMEGA 2560	[3]	Microcontroller
Pesticide tank	[3] [7] [8] [9] [10]	For storing pest
Pixhawk 4	[4] [8]	Microcontroller
Skydroid T12 RC	[4]	Transmitter/receiver
ATMEGA 328	[5]	Microcontroller
ArduPilot	[6]	Microcontroller
2.4 GHz RC transmitter/receiver	[1] [3] [5] [6] [7] [9]	Used for transmit and receive data
Drone stand	[1] [2] [3] [4] [5] [6][7] [8] [9] [10]	Used to place drone above from ground level

(continued)

Table 19.3 (continued)

Reference	Components	Purpose
Fly-sky transmitter/receiver	[10]	Used for transmit and receive data

Table 19.4 UAV in different sector

Sector	Description
Aerial photography	These type of drone are used for aerial view photography and videography such as event, sports, and film making
Shipping and delivery	Drones are useful in delivering lightweight products and food items in cities which have high road traffic
Geographic mapping	Drones are very helpful to capture and analyze colossal areas of coastlines, mountain peaks, and islands
Disaster management	Drones are very helpful in rescuing people when any disasters like fire, tsunami, flood, etc., occur
Precision agriculture	Drones are useful for farmers to monitor huge areas of crop and also helpful for pesticide spraying on crops
Weather forecast	Drones are useful to monitor weather conditions. There are several powerful drones which are sent in hurricanes and tornadoes, so scientists can study its behavior and resistance
Wildlife monitoring	Drones are useful to observe wild animals' behavior and also useful to rescue them during some natural calamities
Law enforcement	Drones are also useful to monitor jostling crowds to ensure public safety and also useful for capturing illegal actions

19.5 UAV in Different Sector

Drone science has been used via protection corporations and tech-savvy buyers for some time. However, the advantages of this science extend well beyond just these sectors. With the rising accessibility of drones, many of the most risky and high-paying jobs inside the commercial zone are ripe for displacement through drone technology. The use cases for safe, within your means options vary from statistics series to delivery. And as autonomy and collision-avoidance applied sciences improve, so too will drones ability to operate increasingly complex tasks. Table 19.4 shown the usage of UAV based on various application domain.

19.6 Research Opportunity and Challenges

Drone technology requires several domains to implement it. Each domain has its own opportunities and limitations while applying into it. Domain-wise research opportunity and challenges identify which shown in Fig. 19.5 and describe each domains

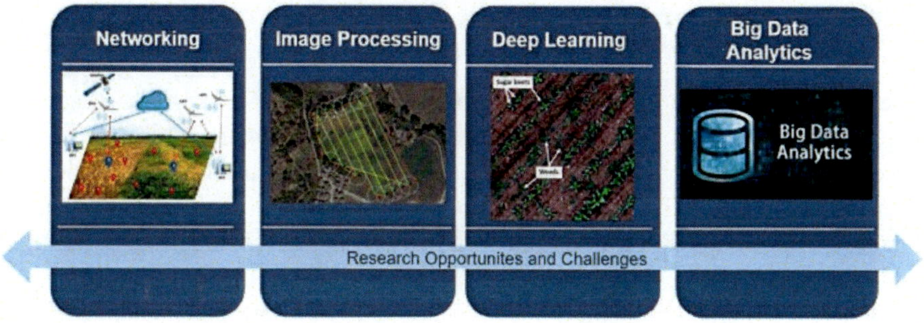

Fig. 19.5 Research opportunities and challenges

with respect to how they are meant to be important in UAV implementation. Each of them defines the roles or their task and how to process and to coordinate with the other domains. Here is an overview of it. From several networking protocols to big data analytics analyzation, it describes the rudimentary form of it.

19.6.1 Networking

ZigBee is a low-speed, low-cost, low-power preferred Wi-Fi transmission technological know-how for use in wireless integrated faraway controllers, farm sprinklers, etc. [20, 21]. The gain of ZigBee potential is affordable solution for low electricity.

Wireless records conversation in private Wi-Fi communication environment, efficiency in strength consumption aspect, implementation of ZigBee in machine ZigBee protocol can be mounted from Bluetooth, IEEE 802.15.3 HR (High Rate)-WPAN [20, 22].

19.6.2 Image Processing

Image processing is a technique to perform some operations on an image, in order to get a greater image or to extract the field image. In agricultural subjects, the use of image processing is higher than guide analysis. The evaluation of crops from drones is faster than normal methods. It is a quicker technique as nicely as accurate. When crop is dead, their leaf modifications are their shade like brown, white, or black. To impervious crops, it is required to discover earlier than they die [20]. So to observe bugs, we can use the insect detection process. Through the lens, it will determine the field area and capture the image [23].

19.6.3 Deep Learning

The addition of deep gaining knowledge of and UAV imagery has proved its adequacy in predicting soil moisture content material irrigation and water administration [20]. Likewise, arbitrary forest, support vector machines, and multiple regression have been used for predicting vegetation production and crop classification. Once the image is captured, it would be sent to the receiver and considering the image as a dataset, and then the decision is taken whether to spray the pesticide or not [23]. Predictive maintenance and monitoring of industrial machine using machine learning helps in diagnose the degradation of drone performance helps to improve the life of drone [24].

19.6.4 Big Data Analytics

Drones can capture, store, and transmit data, giving businesses the chance to integrate greater information into their current processes. The records that drones seize are more superior than in the past. Drones used to acquire simply photographs and videos for visualization. For example, actual estate organizations from time to time use drones to promote the homes they are promoting and attract a wider base of buyers. Drones give fascinating views of the buildings, which can make properties stand out as a pinnacle choice. Today, drones can acquire records about emissions, geodetic facts for land surveyance, and radio signals. They have capabilities that assist all varieties of industries [25, 26].

19.7 Renewing Energy in Drone

As a craving for the greener society is expanding step by step in this day and age, an elective wellspring of energy for drone's is required. There are numerous other elective fuel sources that are available including bio-fuel and hydrogen power devices; however, nothing is just about as boundless when contrasted with sun oriented innovation. Solar energy is one of the limitless accessible sustainable power which can be utilized to expand the perseverance of automated flying vehicle without adding critical mass. The sun-based boards can be ingested in drone which goes about as advantageous force source to expand drone flight time. As the perseverance is expanded, more land region can be covered by the drone.

19.8 Conclusion

In the current decade, technology is very useful in various sectors and precision agriculture is one of them. And the technology is used here to improve crop productivity. The technology used here is really useful in places where living beings' interruptions are not any longer possible for spraying chemicals on crops. It also makes the spraying job faster and less complicated. The device shown here describes the CRP monitoring by using a multispectral camera which is installed in drone. In the current age, to enhance crop production, applied sciences are used in precision agriculture. Instead of humans, those systems are genuinely convenient for spraying chemicals on crops. Indeed, it makes the spraying job much less difficult and easier. By means of the camera connected to the UAV, the proposed system decides the crop survey. The camera captures photographs and analyzes them by means of a geographical indicator in one of the flights and then sprays the pesticides where appropriate. This would maybe even reduce the wasting of water and chemicals. This will also reduce the pest chemicals waste and also water wastage.

19.9 Future Scope

In the future, UAV will be one of the best technologies in each sector and especially in the agriculture applications. By amalgamated UAV with agriculture, several new techniques like image processing, cost reduction, time of flying, batteries with more power, photographic cameras with more pixels, low sprayers, and more nozzle types can formed. An oversized type of experimental research of UAV voluntarily based on remote sensing for agriculture application. There are several extra benefits of these systems in precision agriculture and environmental overlooking.

Acknowledgements This research in this publication was supported by the Nirma University, India, under Minor Funded Research Project scheme with grant number [NU/DRI/MinResPrj IT/2019-20].

References

1. V.K. Suryawanshi, J. Ashok, S.A. Rajmane, S.S. Mali, Design & development of agricultural fertilizer spraying drone with remote controller and autonomous control with low weight aluminium alloy frame structure. J. Remote Sens. GIS Technol. (2019)
2. A.G. Mule, R.P. Chaudhari, Insecticide spraying using Quadcopter, in *International Conference on Intelligent Data Communication Technologies and Internet of Things* (Springer, Cham, 2019), pp. 450–456
3. P. Kardasz, J. Doskocz, M. Hejduk, P. Wiejkut, H. Zarzycki, Drones and possibilities of their using. J. Civ. Environ. Eng. **6**(3), 1–7 (2016)

4. D. Yadav, G. Singh, B.M. Singh, H. Chauhan, K. Gopal, A. Varshney, G. Singh, Quadcopter for agriculture. Int. Res. J. Eng. Technol. **5**(5), 1889–1892 (2018)
5. U.R. Mogili, B.B.V.L. Deepak, Review on application of drone systems in precision agriculture. Procedia Comput. Sci. **133**, 502–509 (2018)
6. T.H.G. Dang, M.T. Vu, Q.H. Vuong, V.T. Nguyen, C.H. Quach, N.T. Truong, M.T. Pham, Design and Implement Low-cost UAV for Agriculture Monitoring (2019)
7. I.D. Pharne, S. Kanase, S. Patwegar, P. Patil, A. Pore, Y. Kadam, Agriculture drone sprayer. Int. J. Recent Trends Eng. Res. (IJRTER) **4**(3) (2018)
8. R. Nivas, T. Naveen, T.C.S. Venkat, T.K.K. Rudra, Automatic spraying of fertilizers/pesticides using drone
9. S.R. Gorantla, Design and modelling of an affordable UAV based pesticide sprayer in agriculture applications, in *2019 Fifth International Conference on Electrical Energy Systems (ICEES)* (IEEE, 2019), pp. 1–4
10. D. Yallappa, M. Veerangouda, D. Maski, V. Palled, M. Bheemanna, Development and evaluation of drone mounted sprayer for pesticide applications to crops, in *2017 IEEE Global Humanitarian Technology Conference (GHTC)* (IEEE, 2017), pp. 1–7
11. S. Reddy, et al., Prototype unmanned aerial sprayer for plant protection in agricultural and horticultural crops, in *Proceedings of the 14th International Conference on Precision Agriculture* (unpaginated, online), International Society of Precision Agriculture, Monticello, IL
12. F. Sarghini, A. De Vivo, Interference analysis of an heavy lift multirotor drone flow field and transported spraying system. Chem. Eng. Trans. **58**, 631–636 (2017)
13. Z. Yanliang, L. Qi, Z. Wei, Design and test of a six-rotor unmanned aerial vehicle (UAV) electrostatic spraying system for crop protection. Int. J. Agric. Biol. Eng. **10**(6), 68–76 (2017)
14. S.S.H. Shah, Development of UAV octocopter based on pesticides spraying system. Univ. Wah J. Sci. Technol. (UWJST) **2**, 13–17 (2018)
15. J. Bendig, A. Bolten, G. Bareth, Introducing a low-cost mini-UAV for thermal- and multispectral-imaging. Int. Arch. Photogramm. Remote Sens. Spat. Inf. Sci. **39**, 345–349 (2012)
16. T. Qing, Z. Ruirui, C. Liping, X. Min, Y. Tongchuan, Z. Bin, Droplets movement and deposition of an eight-rotor agricultural UAV in downwash flow field. Int. J. Agric. Biol. Eng. **10**(3), 47–56 (2017)
17. Y. Huang, W.C. Hoffman, Y. Lan, B.K. Fritz, S.J. Thomson, Development of a low-volume sprayer for an unmanned helicopter. J. Agric. Sci. **7**(1), 148 (2015)
18. D. Giles, R. Billing, Deployment and performance of a UAV for crop spraying. Chem. Eng. Trans. **44**, 307–312 (2015)
19. G. Cai, B.M. Chen, T.H. Lee, An overview on development of miniature unmanned rotorcraft systems. Front. Electr. Electron. Eng. China **5**(1), 1–14 (2010)
20. S.H. Lee, S.H. Yang, Y.M. You, Design and development of agriculture drone battery usage monitoring system using wireless sensor network. Int. J. Adv. Smart Converg. **6**(3), 38–44 (2017)
21. S. Yadav, J. Verma, S. Agrawal, SUTRON: IoT-based industrial/home security and automation system to compete the smarter world. Int. J. Appl. Res. Inf. Technol. Comput. **8**(2), 193–198 (2017)
22. R. Desai, A. Gandhi, S. Agrawal, P. Kathiria, P. Oza, IoT-based home automation with smart fan and AC using node MCU, in *Proceedings of ICRIC 2019* (Springer, Cham, 2020), pp. 197–207
23. A.I. Khan, Y. Al-Mulla, Unmanned aerial vehicle in the machine learning environment. Procedia Comput. Sci. **160**, 46–53 (2019)
24. K.I. Masani, P. Oza, S. Agrawal, Predictive maintenance and monitoring of industrial machine using machine learning. Scalable Comput.: Pract. Exp. **20**(4), 663–668 (2019)
25. S.S. Agrawal, A. Patel, CSG cluster: a collaborative similarity based graph clustering for community detection in complex networks. Int. J. Eng. Adv. Technol. **8**(5), 1682–1687 (2019). https://www.ijeat.org/wp-content/uploads/papers/v8i5/E7475068519.pdf
26. S. Agrawal, A. Patel, Clustering algorithm for community detection in complex network: a comprehensive review. Recent Patents Comput. Sci. **13**(1) (2020)

Author Index

A
Agrawal, Smita, 199
Atri, Amit, 19

B
Bahrani, Prakash, 123

C
Chopade, Sandeep, 175
Chopade, Swati, 175

D
Doshi, Rishabh, 31

F
Fernandez, E., 1, 167

G
Gawade, Sushopti, 175
Gulati, Dhruv, 145
Gupta, Anuj, 9
Gupta, Hemant, 77
Gupta, Kapil, 9

J
Jain, Naveen, 123
Jayachitra, G., 59
Jhamb, Nanak, 145
Joshi, Dheeraj, 135
Jyothi, N. V., 153

K
Kansal, Shivam, 31
Kar, Manoj Kumar, 41
Kathiria, Preeti, 199
Khosla, Anita, 19, 87, 113

M
Mangla, Abhisheak, 145
Maurya, Sanjay, 77

N
Naim, Abdullah, 99
Narasimhan, Abhilash, 69

P
Prajapati, Sandhya, 167

R
Rana, Kailash, 135
Rawal, Vishwam, 199

S
Saini, Rishabh Dev, 31
Saroha, Sumit, 9
Sharma, Rakesh, 113
Singh, A. K., 41
Singh, Arun, 87

A. Khosla and M. Aggarwal (eds.), *Smart Structures in Energy Infrastructure*,
Studies in Infrastructure and Control,
https://doi.org/10.1007/978-981-16-4744-4

Sinha, Ruma, 49, 59
Soni, Yuvraj Praveen, 1, 167
Suresh, K. V., 153

T
Tripathi, Shashi Shekhar, 41

V
Vashist, Devendra, 99, 145
Vidya, H. A., 49, 59
Vinayaka, K. U., 153
Vyas, Trushit, 199

Y
Yadav, Arvind, 77